# The Chemistry of Free Radicals

# 游離基化學槩論

王順福
黃嚴松
吳瑞福
合著

倫敦亞諾出版社印行

# The Chemistry of Free Radicals

Rayson L. Huang

B.Sc., Hon. D.Sc. (Hong Kong), D.Phil., D.Sc. (Oxon)
D.Sc. (Malaya), F.R.I.C., F.R.S.A., F.W.A.
Vice Chancellor, University of Hong Kong, Hong Kong

S. H. Goh

M.Sc. (Malaya), Ph.D. (Chicago), A.R.I.C.
Lecturer in Chemistry, University of Malaya, Kuala Lumpur, Malaysia

S. H. Ong

M.Sc. (Malaya), Ph.D. (Lond.), F.R.I.C.
Professor of Physical Organic Chemistry,
University of Science of Malaysia, Penang, Malaysia

EDWARD ARNOLD

First published 1974 by
Edward Arnold (Publishers) Ltd.
25 Hill Street,
London WIX 8LL

Boards Edition ISBN: 0 7131 2417 2
Paper Edition ISBN: 0 7131 2418 0



Printed in Great Britain by
J. W. Arrowsmith Ltd., Bristol

# Preface

This book aims at giving a broad, comprehensive coverage of the fundamentals of the organic chemistry of free radicals in solution, at a level suitable for the advanced undergraduate reading for an honours degree in chemistry. At the same time, an attempt is made to bring the reader to the forefront of active investigation in a few areas in this vast and exciting field by inclusion of some material of current interest. Care is taken, however, not to overwhelm the student with too much factual information but for those interested in further details relevant references are given.

As is the current trend in organic chemistry texts the book is dominated by a mechanistic approach, and will bridge the gap between standard undergraduate textbooks on organic chemistry and the specialized treatises on free radicals. By placing emphasis on modern theory and its application to the understanding and prediction of the properties and reactions of radicals it is hoped to provide a rational framework for the whole encyclopedia of available factual data. A substantial amount of such data chosen to illustrate the basic principles enunciated are drawn from the authors' own work and an apology is made to those workers in this field whose equally important contributions the authors have not found possible to accommodate in a volume of this size.

The text opens with an introduction dealing with the historical background to the field of radical chemistry, this being followed by a brief description of electron spin resonance and nuclear magnetic resonance spectroscopy. Portrayal in some detail is then given (Chapter 3) of the characteristics of radicals reactions and the concepts underlying the theory of such reactions, including bond dissociation energies, the transition state, the Hammett equation, kinetic isotope effects, kinetics and, briefly, orbital symmetry. Once understanding of these has been achieved, it is hoped the student will find factual data falling into their logical places. After conformation, methods of radical production and relevant aspects of photochemistry have been described, a comprehensive treatment is given of the factors governing radical reactions (Chapter 9). The remaining chapters (10–17) are devoted to a survey of typical radical reactions, with a relatively detailed coverage of addition reactions (Chapter 12), which have extensive application in preparative organic chemistry. The vast field of polymerization, however, has been accorded only cursory treatment.

The authors are grateful to Dr. R. M. Acheson of Queen's College, Oxford, who read some of the chapters and made helpful suggestions. Comments from readers will similarly be appreciated.

R.L.H., S.H.G., S.H.O.

1973

# Contents

# 1

# Introduction
# The 'New' Organic Chemistry

In the history of Chemistry, free radicals are a relative newcomer. The first authentic free radical was prepared only at the beginning of this century, and as to the part radicals play in chemical reactions recognition did not come until more than thirty years later. Being entities bearing unpaired electrons radicals are, as would be expected, extremely reactive and in general capable of transient existence only. Thus it is no wonder that, although there had been no lack of interest to make radicals and isolate them, no attempts were successful until 1900 when Moses Gomberg generated for the first time his famous triphenylmethyl radical, and produced experimental evidence to substantiate his claim. Some twenty-eight years elapsed before the simple methyl and ethyl radicals were prepared. These were shown to exist for only a fraction of a second.

But the discovery of radicals as chemical entities was one thing, and the recognition of their rôle in chemical reactions was another. After Gomberg's discovery, and subsequent successful attempts by others to synthesize new radicals, mainly those related to triphenylmethyl, free radicals were for many years merely a chemical novelty, and their participation as reacting entities in chemical reactions was not recognized until after 1937, following the publication of a series of papers by M. S. Kharasch and by P. J. Flory in the United States, and a significant review by D. H. Hey and W. A. Waters in England. This recognition in fact marked the stage at which the organic chemist's interest in radicals was shifted from the static to the dynamic aspects, and his eyes opened to the vast realm of a new branch of Organic Chemistry which he aptly christened the 'New' Organic Chemistry.

The 'New' Organic Chemistry was therefore 'new' not so much in the sense that radical reactions had not been encountered before, but rather that the recognition of such reactions as involving free radicals was a new concept. In fact, from the very early days of organic chemistry, a number of radical reactions had already been encountered, though not appreciated as such. For instance, the pyrolytic process, a technique much in fashion in the old days, produces radicals. This was at one time a standard method for the degradation of chemical compounds, and almost all natural products had at one time or another suffered this brutal treatment at the hands of the early chemists. Other reactions involving radicals had been encountered for a long time, notably that large group of reactions initiated by sunlight, oxidative processes, electrolytic reactions, the 'drying' of paints and varnishes, and so on. A number of these, in fact, were better known to the chemist as a source of nuisance rather than as chemical processes of any theoretical or practical interest. Among these the deterioration of chemicals by autoxidation perhaps deserves special mention, as it has impact on the organic chemist not only in the research laboratory but also in industry. A bottle of aniline, if no special precautions are taken to protect it from light and air, soon turns brown, then green and black, while a specimen of benzaldehyde soon forms a crust of benzoic acid. Chloroform, if not stabilized by addition of ethanol, is soon oxidized to the poisonous phosgene. Diethyl ether, an indispensable solvent, forms a highly explosive peroxide which has been the cause of many laboratory explosions.

However, although an occasional explosion in the laboratory might be a nuisance to the chemist, what irked him more must have been the apparently wayward and unpredictable manner in which a number of well studied and supposedly well understood chemical reactions took place. These reactions, free radical in nature but not so known to the chemist earlier, completely baffled him when it came to fitting them into the prevailing theory. It will be recalled that, for a long time past, the popular impression of organic chemistry, among practitioners of other branches of scientific disciplines, had been that it was very much of an empirical science comprising a host of facts without theoretical foundation, and that to be successful in organic chemistry all one needed was a colossal memory and immunity towards evil smells and smokes. This picture had begun to change since the early part of the present century, with the evolution of a general theory of organic reactions, the Electronic Theory of Organic Chemistry, to which are attached as pioneers the great names of A. Lapworth, Sir Robert Robinson, and Sir Christopher Ingold. Already in the late 1920's the electronic theory had reached an advanced state of development and was increasingly able to correlate, to a satisfactory degree, much of the information amassed by generations of chemists in the past. The organic chemist was justifiably proud of this achievement, and perhaps a little complacent in being able to explain, in terms of his theory, a good many observations on chemical behaviour and in cases even to predict the course of events in chemical reactions. There were, however, to be unaccommodating exceptions to the established rules. These 'black sheep' among chemical reactions, if we may call them such, seemed to violate all the rules and were none other than what we now know as free radical reactions. As the great majority of reactions known to the earlier organic chemists were of the ionic type, the development of the electronic theory had been based on observations of these reactions. It is no wonder then that the theory, in its state of development at that time, was not adequate to deal with reactions in which a very different reacting species took part.

The classical example among these so-called 'abnormal' reactions, which is of both theoretical and historical interest, is the addition of hydrogen bromide to olefins. When hydrogen bromide adds to an unsymmetrical olefin two modes of addition are possible, and the course taken is prescribed by a rule known as Markownikov's rule which had been enunciated many decades ago. Unfortunately, as time went on a disturbingly large number of exceptions were found to this rule, and what was worse, to what extent the rule was obeyed, or whether it was obeyed at all, appeared to vary from experiment to experiment and from experimenter to experimenter. Obviously there was something very wrong with the rule, and this confusing state of affairs led Kharasch and his co-workers at the University of Chicago to undertake an extensive investigation lasting a number of years. After running over 500 experiments on one single reaction, namely, the addition of hydrogen bromide to allyl bromide, Kharasch made his first breakthrough in discovering what is known as the 'peroxide effect', which accounted for the 'anti-Markownikov' addition of hydrogen bromide, and a little later put forward a mechanism for the reaction which is now generally accepted. The gist of his proposal is this: Markownikov's rule is not obeyed when the reaction takes a non-ionic course, and this easily happens if free radicals have inadvertently been introduced into the reaction system through impure reactants, or in some cases even by contact with air. This 'abnormal' course can, in fact, be made the main reaction by the deliberate introduction of free radicals into the reaction system, e.g. by the addition of a peroxide. There is therefore nothing so abnormal about it after all.

This recognition of free radicals as participants in chemical reactions had far reaching consequences in organic chemistry. For one thing, it uncovered a virgin field for investigation, and the search for new reactions involving radicals was soon on. For another, and this is perhaps of greater fundamental importance, it opened up a new vista in chemical theory. Radicals, after all, arise from one of the two fundamental ways in which the covalent bond can be ruptured and so, whereas hitherto the chemist only concerned himself with ionic reagents when he studied the behaviour of chemical compounds, his study would now be only half complete if he did not examine the effect of free radicals as well.

This interdependence of theoretical development and experimental findings has been encountered time and again before. One might recall, for example, how in the late 19th century aromatic chemistry, although brought into prominence by the coal tar industry, really owed its continued proliferation and its orderly growth to the development of the theory of aromatic substitution. In the same way, with the chemistry of free radicals, the large majority of achievements and the most significant ones came only after, and as a result of, the laying of the theoretical foundation, and this may be considered to have been done in about 1937. Since that time the 'New' Organic Chemistry has gone from strength to strength and, although like aromatic chemistry it also owed much by way of stimulus to an industry (in this case the petroleum industry), it was the sound theory soon to emerge which correlated experimental findings and showed the way for further exploration.

By now we can survey the achievements of the past quarter of a century with some measure of satisfaction. Among these achievements, the most prominent, made during the 1940's, are no doubt the great advances in the chemistry of petroleum, the constituents of which are what organic chemists classify as the paraffin hydrocarbons. To the 19th century chemists these hydrocarbons appeared to be so devoid of chemical activity that they were named 'paraffins' or being interpreted, 'of little affinity'. These earlier practitioners, of course, used only ionic reagents (acids, bases and so on) on these compounds. Quite contrary to the name given them, these hydrocarbons in fact react with great avidity if they are attacked by free radical reagents, and undergo self sustaining reactions of a chain nature which is particularly characteristic of radical reactions in general. Thus paraffin chemistry, of which the sum total of our knowledge forty years ago could be written in a chapter, has now accumulated such a wealth of information that it practically forms a new chemistry by itself. Most of the reactions found to take place with this group of chemical compounds are applicable to organic compounds in general, and the study of the mechanism of these reactions has brought about a new perspective in the theory of organic reactions.

To give a few examples of these radical reactions, and especially those which have useful applications in everyday life, one must mention the study of the chemical process known as combustion, say of petrol or other fuels, and the consequent development of 'antiknocks' and other additives to aid the process of combustion and hence make better fuels. From a study of the process known as autoxidation we have come to understand how fats and oils, and certain other foodstuffs, deteriorate and have developed 'antioxidants' or 'inhibitors' which retard these chemical changes and can be used as food preservatives. Incidentally, this same type of reaction is responsible for the decomposition of the chemical reagents mentioned earlier, namely, aniline, benzaldehyde, and ether and indeed, for the 'ageing' or deteriorating of unprocessed natural rubber, all of which substances can be stabilized to varying degrees by the addition of small quantities of inhibitors. While autoxidation is often a nuisance and special measures have to be taken to check it, it has been put to good use in several instances. For example the aerial oxidation of cumene (isopropylbenzene), which is carried out on an industrial scale, produces the antiseptic phenol and acetone. The drying of paints and varnishes is also a process involving free radicals, and research in this field has led to new materials designed to suit specific requirements. Again, by means of the so-called 'cracking' process, i.e., pyrolysis under controlled conditions, the relatively useless larger hydrocarbon molecules present in petroleum are broken into smaller molecules suitable for use as motor fuels. In this way we manage to double the yield of fuels obtainable from crude petroleum. Conversely, the process of polymerization, by which small molecules are built up into giant ones according to specifications, gives us 'tailored' molecules of desired sizes and shapes to serve a variety of purposes.

Mention must also be made of that prominent group of radical reactions, taking place in the vapour or liquid phase, initiated by irradiation with sunlight, ultraviolet, or gamma rays, all of which bring about the breaking of covalent bonds to generate radicals. Photochemistry, in particular, has become one of the most rapidly developing areas of research. Recent photochemical studies have thrown much light on the nature and reactions of chemical bonds.

This resumé of radical reactions would not be complete without a brief reference to the rôle of radicals in reactions taking place in biological systems. This was the subject of a good deal of speculation for a long time but the study of which had not gained much headway until recently, especially since the development of the electron spin resonance technique some years ago. The importance of the problem is evident from the list of biological processes in which free radicals have been shown to be active participants by this new technique. These include enzymatic oxidation and reduction, photosynthesis, chemiluminescence, damage to living cells by irradiation and initiation of malignant growth. Furthermore, the technique of spin labelling, e.g., by the use of stable nitroxide radicals, promises to give greater insight into cellular structure and function.

Finally, mention must be made of the part which modern instrumentation has played in the unravelling of the intricacies in radical reaction mechanisms. Such powerful tools as gas chromatography, mass spectrometry, nuclear magnetic and electron spin resonance have made possible detailed and refined analysis undreamt of twenty years ago. New avenues of research have thereby continually been opened up, and the flood of new knowledge emerging continues unabated.

# 2
# Detection of Radicals

Except for the naturally occurring free radicals (e.g., $O_2$ or NO) and those considerably stabilized by resonance, radicals are in general very reactive entities and capable of transient existence only. Nevertheless a number of methods have been devised for their detection and identification. These encompass chemical means of trapping and isolation, as well as direct detection by spectroscopic methods.

## 2.1 Chemical methods

Radical intermediates in chemical reactions are usually of such high reactivity as to preclude their isolation or even detection by physical methods. Their transient existence may nevertheless be ascertained by trapping them with reagents for which they have high affinity and with which they combine to give stable products which can then be isolated and identified. The involvement of a carbene intermediate can, for example, be inferred if in the presence of an olefin, added as a trapping agent, cyclopropanes are produced. Similarly, the addition of toluene to a radical reaction will allow intermediate radicals to abstract α-hydrogen atoms to generate relatively stable benzyl radicals, which will persist in the reaction system until they dimerize to yield bibenzyl, a readily identifiable product. If no such trapping product is found the involvement of an intermediate radical would appear less likely but nevertheless cannot be ruled out.

Quite frequently the nature of the products from a chemical reaction offers a clue as to whether the reaction involves radicals or not. Homolytic reactions generally give rise to a variety of products arising from abstraction, addition, dimerization, fragmentation and other processes, and through a careful analysis of the products the radicals involved can often be identified.

The homolytic nature of a reaction may also be inferred from evidence of catalysis or inhibition. If the addition of radical producing substances (e.g., peroxides) or the application of ultraviolet light, which gives rise to radicals through homolytic bond cleavage, greatly accelerates the reaction it is likely that a free radical mechanism is in operation. Conversely, the addition of inhibitors which reduce the concentration of reacting radicals will slow down or inhibit the reaction. Such inhibitors are usually substances containing active hydrogen atoms (e.g., phenols and aromatic amines) and these render the chain-propagating radicals unreactive through hydrogen donation to form inhibitor radicals which are too stable to react further or else take up some new reaction pathway. In either case the inhibitor slows down or even stops the original radical reaction. Sometimes the trapping of transient intermediates by scavengers gives stable radicals which can easily be detected by electron spin resonance techniques.

Stable paramagnetic species, e.g., oxygen and nitric oxide, can also inhibit radical reactions by reacting with the chain-propagating species. Oxygen is well-known for its unpredictable effect on radical reactions: small quantities of oxygen may initiate a reaction whereas large quantities may completely inhibit it. In many cases it is directly responsible for the induction period of a reaction (see also autoxidation, Section 10.4).

## 2.2 Electron spin resonance

The electron spin resonance (e.s.r.)[1] method for the detection and study of radicals has in recent years achieved great prominence and will be described in some detail here.

### (A) *Theory*

The electron has spin and an associated magnetic moment and can be considered as a small magnet. In the presence of an external magnetic field it will align itself parallel or anti-parallel (in opposition) to the field according to quantum restrictions. Thus the energy level of the electron in an external magnetic field is split into two possible values, the difference being given by $\Delta E$ (see Fig. 2.1). A resonance condition may be achieved if electromagnetic radiation of the correct energy is applied and excitation from the lower energy state to the upper one will give rise to an absorption spectrum. The equations for the resonance condition are given below

$$\Delta E = g\left(\frac{ehH}{4\pi mc}\right) = g\beta H$$

$$\Delta E = h\nu = g\beta H$$

where $g$ is a constant of approximately 2, $e$ the electronic charge, $h$ Planck's constant, $m$ the mass of the electron, $c$ the velocity of light, $H$ the applied magnetic field, $\beta$ the Bohr magneton and $\nu$ the frequency of the applied electromagnetic radiation. Attention should here be drawn to the similarity

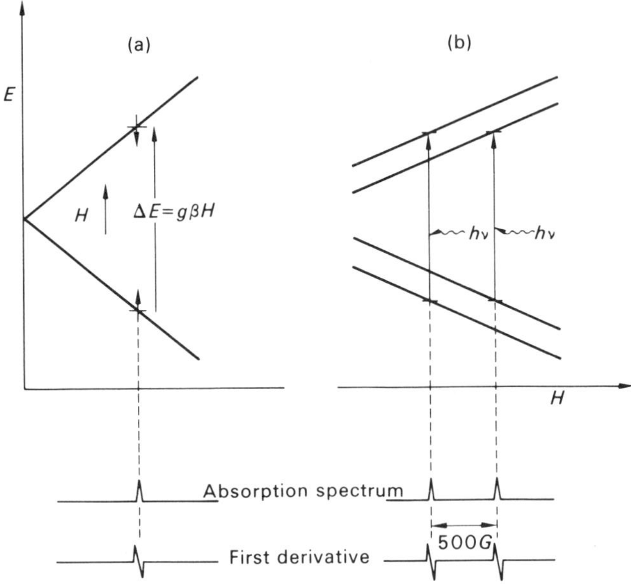


**Fig. 2.1** (a) Diagram showing the transition and spectra of a free electron in a magnetic field; (b) diagram showing transitions and spectra of hydrogen atom.

of the theory of electron spin resonance (sometimes referred to as electron paramagnetic resonance (e.p.r.)) to that of nuclear magnetic resonance. Under practical conditions $\nu$ is fixed at about 9000 MHz while $H$ is varied and resonance occurs at about 3200 gauss; $\Delta E$ is approximately 1 cal mole$^{-1}$. There is a very slight excess in population of the lower energy state over that of the upper one in accordance with the Boltzmann distribution so that an absorption spectrum can be obtained. Fig. 2.1 shows diagrammatically the energy levels and spectra of a free electron and of a hydrogen atom. For the hydrogen atom the electron interacts not only with the external magnetic field but also with the nucleus which has a spin of $\frac{1}{2}$. This interaction leads to a splitting of the energy levels as shown; the coupling constant of 500 gauss is obtained from the e.s.r. spectrum and this measures the extent of the electron-nuclear interaction.

In an organic radical the number of absorption lines in the e.s.r. spectrum (i.e., hyperfine structure) can be many as the unpaired electron can usually interact with a large number of protons of the radical. The spectrum of the methyl radical, with four hyperfine lines resulting from electron interaction with three equivalent protons, is shown in Fig. 2.2($a$), from which the similarity of the splitting pattern of the e.s.r. spectrum and that of nuclear magnetic resonance should be noted. The splitting in the methyl radical

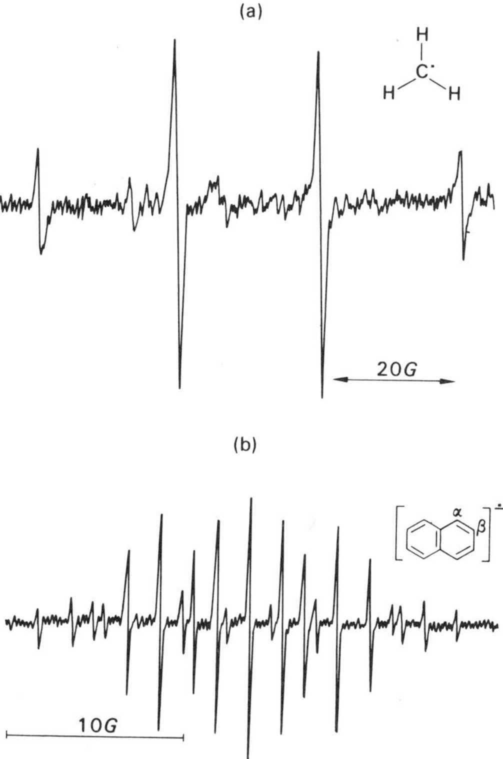


**Fig. 2.2** (a) Spectrum of methyl radical in aqueous solution (R. O. C. NORMAN and B. C. GILBERT, *Adv. Phys. Org. Chem.*, 1967, **5**, 57); (b) Naphthalene radical anion (A. R. FORRESTER, J. M. HAY, and R. T. THOMSON, *Organic Chemistry of Stable Free Radicals*, Academic Press, 1968, p. 93; K. MARKAU and W. MAIER, *Z. Naturf.*, 1961, **16a**, 636.)

spectrum is smaller than that of the hydrogen atom because in the former the protons are much further away from the unpaired electron resulting in less interaction. A more complicated spectrum, that of the naphthalene radical anion, is shown in Fig. 2.2(*b*). Here there are four α- and four β-protons which would give $(4 + 1) \times (4 + 1)$ or 25 lines, all of which have been observed experimentally. Hyperfine splittings for this radical have been observed to be 4.95 gauss and 1.87 gauss, arising from coupling to the α- and the β-protons respectively.

The experimental observables for a radical are the $g$ value, hyperfine structure and line shape. The unpaired electron is distributed over the atoms of the whole molecule and this distribution is described in terms of the electron spin density. The McConnell equation[2]

$$a_i = Q\rho_i$$

gives the relationship of the spin density, $\rho_i$, at the carbon atom $i$ with hyperfine splitting of $a_i$ gauss and $Q$ is a constant of about $-23$ gauss. For the methyl radical, $a$ has been found to be $-23$ gauss and $\rho$ equals 1. The consecutive introduction of one, two, and three benzene rings into the methyl radical, to give the radicals $Ph\dot{C}H_2$, $Ph_2\dot{C}H$, and $Ph_3\dot{C}$, gradually reduces the spin density of the unpaired electron at the methyl carbon from 1 to 0.72, 0.66, and 0.58, respectively. These values clearly demonstrate the effect of delocalization of the unpaired electron by the phenyl groups. In the case of the free benzyl radical delocalization of the unpaired electron as shown by the magnitude of the hyperfine splittings of 5.1, 1.6, and 6.3 gauss for the *ortho*, *meta*, and *para* protons, respectively, is directed mainly to the *ortho* and *para* positions, and this is again as expected theoretically. For the naphthalene radical ion which has hyperfine splittings of 4.95 gauss and 1.87 gauss at the $\alpha$- and $\beta$-carbon atoms, the spin densities are approximately 0.2 and 0.07, respectively. This indicates higher reactivity at the $\alpha$-position as has been found experimentally.

### (B) *Applications*

The greatest advantage of the e.s.r. method for detection of free radicals lies in its high sensitivity; concentrations of as low as $10^{-10}$ M of radicals may be detected under favourable conditions. While detection of intrinsically stable radicals will present no problems, short-lived entities and transient radical intermediates in chemical reactions may be detected only if special techniques are used. One such innovation, referred to as the matrix isolation technique, consists in generating the unstable radical in a suitable inert matrix. In this way entities including triplet carbenes have been generated by irradiation at low temperatures of the appropriate precursors dissolved in a transparent solid solvent, in which rigid medium these reactive species are rendered immobile and may be studied at leisure by the e.s.r. method. Short-lived radicals can also be generated for e.s.r. spectroscopy by continuous electrolysis or photolysis, while for the detection and identification of radical intermediates in mechanism studies Waters[2] has developed a continuous flow system in which radicals produced immediately after mixing of reactants can be observed by e.s.r.

The e.s.r. method not only makes possible the detection of radicals but also provides a useful tool in the study of their electronic structures and correlations with chemical structure, properties, and molecular orbitals.[3] It can also be applied to investigations on kinetics including measurement of the rates of radical formation and destruction, lifetimes of radicals and the nature of their subsequent conversions.

One interesting application is found in the study of the relationship between structure and conformation.[4] For example, the vinyl radicals (1) studied at low temperatures have been shown to consist of interconverting forms. At 4.2 K the e.s.r. spectrum of (1*a*) shows that the $\beta$-hydrogens are non-equivalent but at 90 K the spectrum is simpler, an observation accounted for by an equilibrium of interconverting forms with the odd electron in an $sp^2$ orbital rather than a $p$ orbital, as shown. The methyl substituted vinyl radical (1*b*) shows similar behaviour but the barrier to inversion is higher than that in the unsubstituted radical. For the cyclopropylcarbinyl radical e.s.r. studies have also revealed that there is a preferred conformation, the semidione radical ion (2) exhibiting high preference for the 'bisected' conformation shown below.

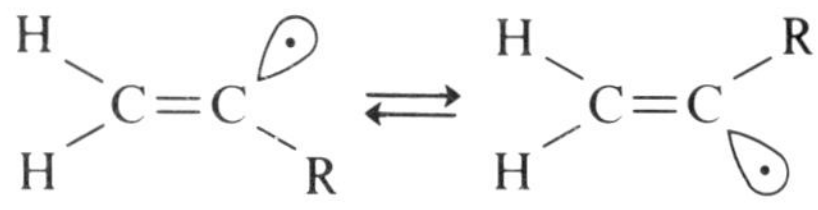

(1)

(1*a*) R = H, inversion barrier 2 kcal mole$^{-1}$.
(1*b*) R = Me, inversion barrier 3.3 kcal mole$^{-1}$.

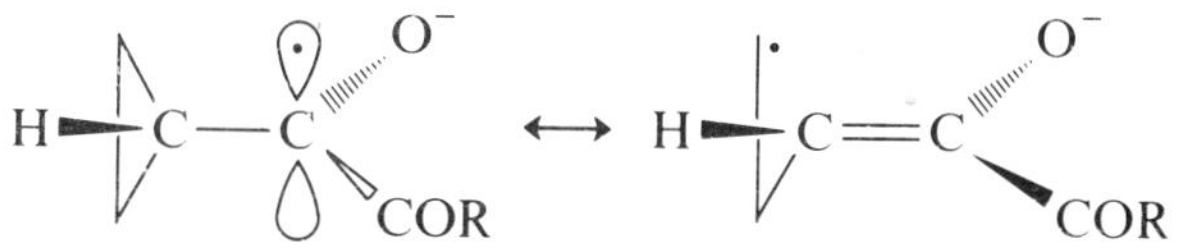

(2)

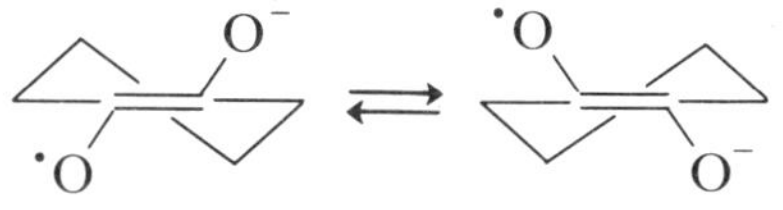

(3)

The interconversion of half-chair conformers of the six-membered rings of the semidione radical anions (3) has been revealed from a study of their temperature dependent e.s.r. spectra. Again, from the spectra of the ketyl radical anions derived from cyclohexanone and its 4-*t*-butyl derivative, it has been concluded that, at room temperature, the former is rapidly interconverting between two chair conformations whereas the latter is essentially fixed in one conformation. The existence of rotational energy barriers can also be discerned from e.s.r. splittings and hindered rotation about double bonds has indeed been inferred in several radicals.

A relatively recent application of e.s.r. spectroscopy is found in the use of spin labels[5] to probe the structure and function of biological macromolecules. Relatively stable nitroxide radicals are useful as spin labels as they can be attached to a biological macromolecule by reaction with a chemically reactive functional group present in the molecule, while the radical centre remains chemically unaffected. Typical spin labels are (4) and (5) indicated below, but other nitroxide radicals of greater complexity are available. Proteins can be 'tagged' with such spin labels at the active sites or at other

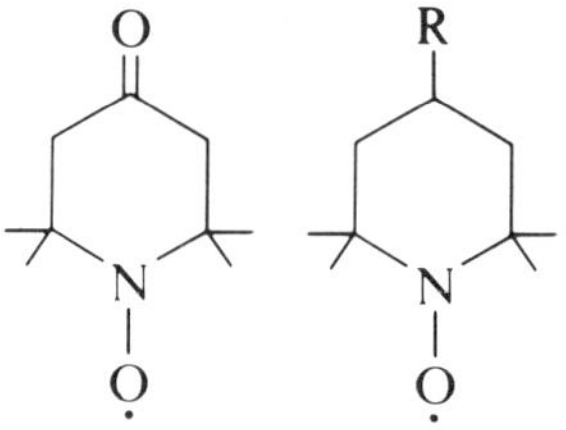

(4) (5) R = OH or $NH_2$

well defined positions, and the e.s.r. signal studied. The advantage of this over other methods lies in the sensitivity of the e.s.r. signal to the local environment and in the fact that rapid molecular motion can also be measured. More recently the nitroso compounds (6) and nitrones (7) have also been employed to trap radical intermediates in organic reactions, forming stable nitroxide radicals which can readily be identified by e.s.r.

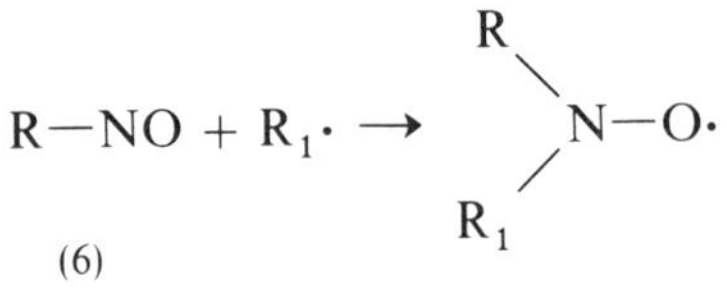

$$\underset{(6)}{R{-}NO} + R_1\cdot \rightarrow R(R_1)N{-}O\cdot$$

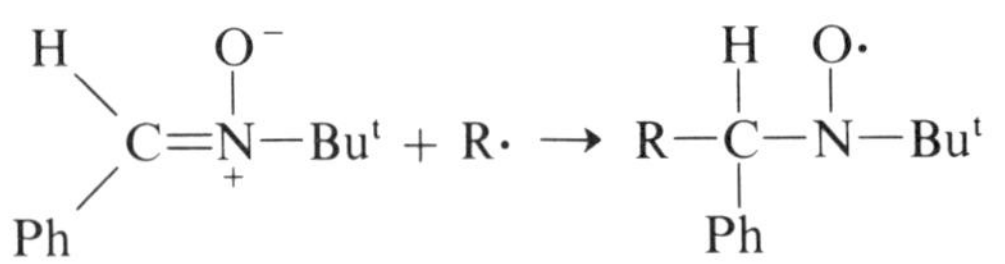

$$\underset{(7)}{H(Ph)C{=}\overset{+}{N}(O^-){-}Bu^t} + R\cdot \rightarrow R{-}CH(Ph){-}N(O\cdot){-}Bu^t$$

(C) *Triplet molecules*

A species containing two unpaired electrons, i.e., a triplet or diradical, has three spin states characterized by $M_s = 1$, 0, and $-1$, on which the effects of a magnetic field are shown in Fig. 2.3. Transitions may be observed

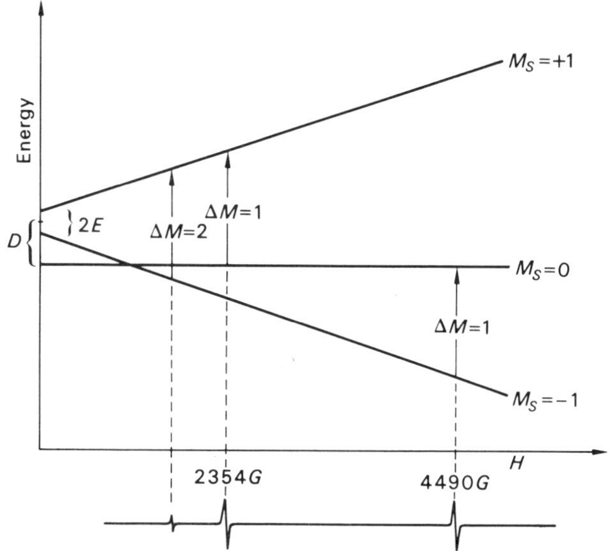


**Fig. 2.3** Energy levels of a triplet as a function of magnetic field. The e.s.r. spectrum shown diagrammatically is that for a naphthalene triplet with $H$ parallel to the $z$ axis using a frequency of 9654 MHz.

according to the selection rule $\Delta M = \pm 1$ and are often anisotropic, i.e., they vary with the orientation of the species in the applied magnetic field. Randomly oriented triplets in solution have the spectrum spread out over so wide a range of field as to be often undetectable and one method of obtaining the e.s.r. spectrum of such species is to generate them in a 'host' crystal, by irradiation, so as to ensure well defined orientations with respect to the 'host' crystal lattice. A simpler method would be to irradiate an appropriate precursor of the triplet in a rigid glass at low temperatures[7] (see matrix isolation technique mentioned earlier). The randomly oriented triplets produced in such a sample will give e.s.r. absorptions when a molecular axis of the triplet species is approximately parallel to the applied magnetic field.

The interaction of the two unpaired electrons in the triplet molecule is in fact appreciable and is present even without application of a magnetic field. This interaction is characterized by two parameters, $D$ and $E$, referred to as zero-field parameters, the former being a measure of the spin interaction along the molecular axis ($z$ axis) while the latter refers to the spin interaction along the $x$ and $y$ axes. The zero-field parameters of a number of triplet species are given in Table 2.1, from which it can be seen that in general spin-spin interactions become small, i.e., $D$ is small, when the two spins are well separated over a conjugated system.

**Table 2.1** Zero-field Parameters of some Triplet Species

| Triplet | Structure | $D$ (in $cm^{-1}$) |
|---|---|---|
| Methylene | $H_2\dot{C}\cdot$ | 1.017 |
| Phenylnitrene | $Ph\dot{N}:$ | 1.00 |
| Phenylcarbene | $Ph—\dot{C}H$ | 0.515 |
| 4-Nitrophenylcarbene | 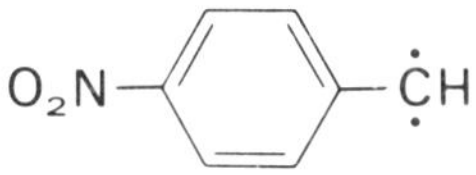 | 0.480 |
| Diphenylcarbene | 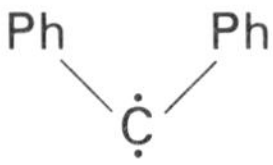 | 0.406 |
| Naphthalene | 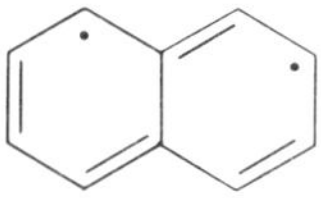 | 0.1003 |
| 4-Phenylene-bis-phenylmethylene | 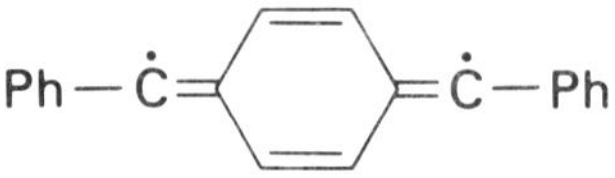 | 0.0521 |

E.s.r. data has also provided information on the structure of carbenes, the bent structure of triplet arylcarbenes (see Chapter 17), for example, having been confirmed by this method.

## 2.3 Chemically induced dynamic nuclear polarization

During rapid radical reactions, surprisingly strong emission and/or absorption signals may be observed in the nuclear magnetic resonance spectra. This phenomenon was experimentally observed by Fischer (1967)[8a] and independently by Ward (1967)[8b] and has been referred to as Chemically Induced Dynamic Nuclear Polarization (CIDNP). An elegant theoretical treatment of this has been provided by Closs[9] and Kaptein and Oosterhoff[9] who realized that radical pairs are directly responsible for the experimental observations. Ordinarily, proton n.m.r. absorption spectra arise due to the slight excess of nuclear spin state population at the lower energy level according to the Boltzmann distribution. The energy diagram is essentially similar to that drawn for e.s.r. (Fig. 2.1). For CIDNP spectra this type of thermal equilibrium population is obviously not valid; the higher energy level may become more highly populated resulting in an emission signal by downward transitions, or the lower level may be more populated than normal, in which case an enhanced absorption spectrum will be observed. Closs[9] explained

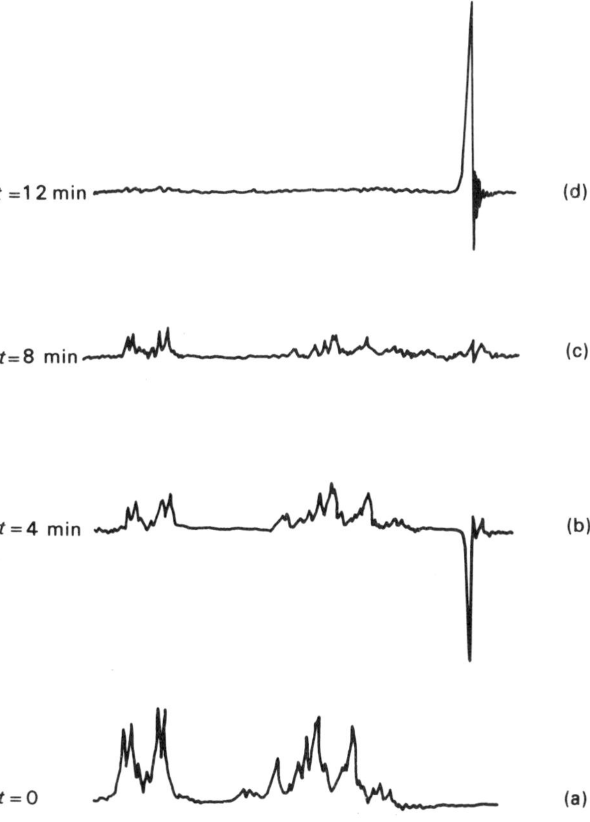


**Fig. 2.4** Changes with time in the N.M.R. spectra of benzoyl peroxide in cyclohexanone at 110 °C (FISCHER and BARGON)[8c]; (a) absorption spectrum of benzoyl peroxide, (b) reduced intensity of benzoyl peroxide protons due to thermal decomposition and emission spectrum of benzene protons formed with non-equilibrium nuclear state populations, (c) appearance of the benzene absorption signal, and (d) normal absorption spectrum of benzene. Reprinted by permission of the American Chemical Society.

that CIDNP behaviour arises from the unique properties of a radical pair in a magnetic field. The electron of the radical, already being coupled to the surrounding nuclei (i.e., e.s.r. hyperfine coupling as discussed in Section 2.2), will experience further interaction with that of the other radical as well as with the magnetic field. Such interactions for pairs of unlike radicals lead to singlet triplet mixing of the nuclear spin states. This means that nuclear spin state populations of the products derived from these radical pairs will vary drastically (i.e., be polarized) from that of normal thermal equilibrium and will consequently give rise to CIDNP signals. Although the singlet triplet mixing may be very small because of the short lifetime ($10^{-10}$–$10^{-9}$ sec) of radical pairs in solution, calculation has shown that this is sufficient to produce the dramatic enhancements of the observed n.m.r. signals. The detailed theoretical treatment[9] which is beyond the scope of this book has provided fairly accurate quantitative as well as qualitative predictions.

The CIDNP method is potentially a powerful tool in mechanistic studies of radical reactions in solution. Since it detects products from radical pairs, it can clearly be used to distinguish radical mechanisms from those of ionic or concerted pathways; its sensitivity in such dynamic processes can indeed be better than that of e.s.r. methods. It is also a far simpler method of detection of radical formation than those using racemization techniques. As an illustration, the spectral changes from the study of the homolytic decomposition of benzoyl peroxide, depicted below, are shown and explained in Fig. 2.4. The benzene emission signal arises from the radical pair (8).

$$\mathrm{PhC(=O)-O-O-C(=O)Ph} \longrightarrow (\mathrm{PhC(=O)-O^{\bullet}\ \ {}^{\bullet}O-C(=O)Ph}) \longrightarrow \underset{(8)}{(\mathrm{Ph^{\bullet}\ \ {}^{\bullet}O-C(=O)Ph})}$$

$$(8) \longrightarrow \mathrm{Ph^{\bullet}}$$

$$\mathrm{Ph^{\bullet} + RH(solvent) \longrightarrow PhH + R^{\bullet}}$$

The radical nature of the reaction between n-butyl bromide and n-butyl lithium has similarly been detected by CIDNP. First examples of strong nuclear polarizations have also been observed by Closs[9] in photochemically initiated reactions proceeding through triplet state intermediates. For instance, the photolysis of diphenyldiazomethane, $Ph_2CN_2$, in toluene gives rise to a radical pair (9), singlet or triplet, which finally leads to a singlet 1,1,2-triphenylethane. In accord with theory, singlet-triplet mixing in such radical pairs leads to large nuclear polarizations in the products derived from radical coupling.

$$Ph_2CN_2 \xrightarrow{h\nu} Ph_2\dot{C}^{\cdot} + N_2$$

$$Ph_2\dot{C}^{\cdot} + PhCH_3 \longrightarrow (Ph_2CH^{\cdot}\ \ ^{\cdot}CH_2Ph) \longrightarrow Ph_2CH{-}CH_2Ph$$

(9)

## 2.4 Other methods

Radicals can also be detected and studied by means of electronic spectroscopy, i.e., absorption spectroscopy in the ultraviolet and visible region. In a radical the interaction of the unpaired electron with other bonds usually alters its electronic energy levels and this in turn causes shifts in the absorption regions. Strong absorption bands in the visible region are often shown by stable radicals in solution and their concentrations may thus be readily estimated, albeit only roughly since coloured decomposition products frequently interfere with the measurements.

For the determination of relatively high concentrations of stable radicals two other methods, the cryoscopic and the magnetic susceptibility methods, have long been known. Both were employed in the earlier studies of triarylmethyl radicals and their dissociation from what were previously believed to be hexa-arylethanes. The cryoscopic method is essentially a molecular weight determination: the apparent molecular weight obtained for a sample of triarylmethyl radicals in equilibrium with the 'dimer' allows the percentage of the free radical in solution to be computed. The magnetic susceptibility method depends on the paramagnetic property (i.e., attraction to a magnetic field) of a free radical in contrast to normal compounds which, with all the electrons paired, are diamagnetic (repulsive to a magnetic field). Measuring the magnetic susceptibility of a sample, usually done with a sensitive Gouy balance, thus ascertains the concentration of radicals. Both methods, however, give results which are only approximate and at times erratic, in a few cases differing, for the same radical, by as much as an order of magnitude. Such inconsistencies[10] have been attributed to various causes. One of these is the presence of impurities in samples. Another source of error inherent in both methods is the relatively long time required for the measurement to be taken, often as long as two hours. As most 'stable' radicals are in fact not that stable but will undergo isomerization and disproportionation during the course of the experiment, both molecular weight and magnetic measurements become subject to substantial inaccuracies.

While it is possible to estimate the concentrations of stable or long-lived radicals by electronic spectroscopy or by cryoscopic or magnetic susceptibility measurements it should be noted that these methods cannot be used for the detection of short-lived radicals. The e.s.r. method discussed earlier is far more accurate and versatile and can readily detect even low concentrations of both stable and unstable radicals.

Of historical interest only is the method of Paneth (1929) which will be briefly mentioned here. Tetramethyllead when heated in a glass tube deposits a mirror of metallic lead with the production of ethane gas, formed from coupling of free methyl radicals produced by pyrolysis of the organolead compound. The lead mirror can be removed by passing over it a stream of hot tetramethyllead vapours, as these contain methyl radicals which react with the lead mirror to regenerate tetramethyllead. By identifying the organometallic product obtained, the identity of the radical can be ascertained. Mirrors of metals other than lead, such as zinc and antimony, can also be used.

## References


1*a* M. Berson and J. C. Baird, *An Introduction to Electron Paramagnetic Resonance*, W. A. Benjamin, New York, 1966;
*b* M. C. R. Symons, *Adv. Phys. Org. Chem.*, 1963, **1**, 283.
2 W. A. Waters, *Organic Reactions Mechanisms*, Special Publication No. 19, The Chemical Society, London, 1965, p. 72.
3*a* E. Muller, A. Rieker, K. Scheffler, and A. Moosmayer, *Angew. Chem., Int. Ed.*, 1966, **5**, 6;
*b* *Radical Ions*, eds. E. T. Kaiser and L. Kevan, Interscience, New York, 1968.
4 R. O. C. Norman and B. C. Gilbert, *Adv. Phys. Org. Chem.*, 1967, **5**, 53.
5 O. H. Griffith and A. S. Waggoner, *Acc. Chem. Res.* 1969, **2**, 17.
6 C. A. Hutchinson, Jr. and B. W. Mangum, *J. chem. Phys.*, 1961, **34**, 908.
7*a* E. Wasserman, L. C. Snyder, and W. A. Yager, *J. chem. Phys.*, 1964, **41**, 1763;
*b* A. M. Trozzolo, *Acc. Chem. Res.*, 1968, **1**, 329.
8*a* J. Bargon and H. Fischer, *Z. Naturf.*, 1967, **22a**, 1551;
*b* H. R. Ward and R. G. Lawler, *J. Am. chem. Soc.*, 1967, **89**, 5518;
*c* H. Fischer and J. Bargon, *Acc. Chem. Res.*, 1969, **2**, 110.
*d* H. R. Ward, *Acc. Chem. Res.*, 1972, **5**, 18.
9*a* G. L. Closs, *J. Am. chem. Soc.*, 1969, **91**, 4552;
*b* R. Kaptein and L. J. Oosterhoff, *Chem. Phys. Lett.*, 1969, **4**, 195, 214;
*c* G. L. Closs in *XXIIIrd* Internat. Congress of Pure and Applied Chem. *1971*, Butterworth, London, 1971, Vol. 4, p. 19;
*d* G. L. Closs and A. D. Trifunac, *J. Am. chem. Soc.*, 1970, **92**, 2183.
10*a* S. T. Bowden, *J. chem. Soc.*, 1957, 4235;
*b* C. S. Marvel, J. F. Kaplan and C. M. Himel, *J. Am. chem. Soc.*, 1941, **63**, 1892.

# 3

# Characteristics of Radical Reactions

Before discussing the characteristics of radical reactions a brief account will be given of the main types of reactions encountered in radical chemistry.

## 3.1 Classification of radical reactions

Reactions of radicals may be broadly grouped under radical combination processes and radical transfer processes; the former involves two radicals and leads to non-radical products in which all the electrons are paired, while the latter results in the generation of a new radical which reacts further with molecules around it, thereby often setting up a self-propagating reaction chain. These are illustrated in the following summary of typical radical reactions.

*Combination or coupling:*

$$R\cdot + R'\cdot \rightarrow R{-}R'$$

Combination of two radicals leads to bond formation and is therefore energetically favourable. The process is known to be extremely fast requiring little or no activation energy, but since in solution the concentrations of radicals are extremely low and reactions between them depend on the square of their concentrations such a process is in general not as important as transfer reactions with the solvent. However, combination reactions may predominate in the case of a relatively stable radical, such as free benzyl, which is not reactive enough to attack other molecules present in the system, and persists in solution until it collides with a like radical and couples with it to yield the 'dimer'. Combination may also take place readily given favourable conditions such as that existing in what has been referred to as the solvent 'cage'. When generated in solution from a radical initiator by bond cleavage, radicals are often formed in pairs in close proximity to each other within a solvent cage, and do not immediately separate. Before diffusing out of the cage they may undergo collision with each other and combine to regenerate the initiator or give other products. This type of reaction has been called a 'cage reaction'.

*Disproportionation:*

$$\mathrm{H{-}\overset{|}{\underset{|}{C}}{-}\overset{|}{\underset{|}{C}}\cdot + H{-}\overset{|}{\underset{|}{C}}{-}\overset{|}{\underset{|}{C}}\cdot \rightarrow H{-}\overset{|}{\underset{|}{C}}{-}\overset{|}{\underset{|}{C}}{-}H + \overset{}{C}{=}\overset{}{C}}$$

The transfer of a $\beta$-hydrogen atom from one radical to another to give non-radical products is known as disproportionation. This process, like that of radical coupling, is energetically favourable, two bonds being formed for one broken. For example, two ethyl radicals can disproportionate to yield ethane and ethylene.

*Redox reactions:*

Just as radicals may be formed by one-electron transfer redox reactions, they may be similarly destroyed. Reduction of radicals by transition metal ions leads to anions while oxidation gives cations, e.g.,

$$\mathrm{HO\cdot + Fe^{II} \rightarrow HO^{-} + Fe^{III}}$$

$$\mathrm{Ar\cdot + Cu^{I} \rightarrow Ar^{+} + Cu^{0}}$$

*Displacement:*

$$\mathrm{R\cdot + A{-}B \rightarrow R{-}A + B\cdot}$$

Displacement or substitution as in ionic displacement ($S_N$ reactions) rarely occurs in radical chemistry except for the abstraction of hydrogen and halogen atoms, if these may also be looked upon as displacements. Homolytic substitution, which may be denoted as $S_H2$, on saturated carbon has not been encountered but instances of such occurrence on hetero-atoms have been reported.

*Addition:*

$$\mathrm{R\cdot + \;C{=}C \rightarrow R{-}\overset{|}{\underset{|}{C}}{-}\overset{|}{\underset{|}{C}}\cdot}$$

The addition of a radical to an unsaturated compound, generally an olefin but occasionally an acetylenic or carbonyl compound as well, usually leads to the formation of another, more stable, radical. If this resulting radical adds to another molecule of the unsaturated substrate the process is referred to as telomerization. Further repeated addition, which may eventually consume hundreds or thousands of substrate units, is also possible and results in radical polymerization. That of styrene, for example, is depicted below.

$$\mathrm{Ph\cdot + CH_2{=}CHPh \rightarrow PhCH_2{-}\dot{C}HPh}$$

$$\mathrm{PhCH_2\dot{C}HPh + CH_2{=}CHPh \rightarrow PhCH_2\underset{\displaystyle Ph}{\underset{|}{C}}H{-}CH_2\dot{C}HPh}$$

$$\mathrm{PhCH_2\underset{\displaystyle Ph}{\underset{|}{C}}HCH_2\dot{C}HPh \rightarrow Ph{-}(CH_2CHPh)_n{-}CH_2\dot{C}HPh}$$

An addition reaction is also involved as the first step in homolytic aromatic substitution.

*Fragmentation:*

This is a process in which a radical breaks down by the scission of a bond $\beta$ to the radical centre resulting in the formation of a smaller radical and an unsaturated molecule. Some examples of radical fragmentation are given below.

$$\mathrm{Me{-}\overset{\displaystyle Me}{\overset{|}{\underset{\displaystyle Me}{\underset{|}{C}}}}{-}O\cdot \rightarrow Me_2C{=}O + Me\cdot}$$

$$\mathrm{PhCH_2O{-}\underset{\displaystyle H}{\underset{|}{\dot{C}}}{-}Ph \rightarrow PhCH_2\cdot + O{=}\underset{\displaystyle H}{\underset{|}{C}}{-}Ph}$$

$$\mathrm{Ph{-}C(=O){-}O\cdot \rightarrow Ph\cdot + CO_2}$$

Two special cases of fragmentation may also be mentioned here, namely, the radical 'unzipping' of a polymer and the decomposition of the 4-nitrocumyl chloride radical anion, shown as follows:

$$\mathrm{RCF_2CF_2CF_2CF_2\cdot \rightarrow R\cdot + 2CF_2{=}CF_2}$$

$$\mathrm{[4\text{-}O_2N{-}C_6H_4{-}CMe_2Cl]^{\dot{-}} \rightarrow 4\text{-}O_2N{-}C_6H_4{-}\dot{C}Me_2 + Cl^-}$$

*Insertion:*

This reaction is peculiar to carbenes and nitrenes. The insertion of methylene, $H_2C$:, into the C—H bonds of pentane occurs indiscriminately to give isomeric hexanes:

$$CH_3{-}(CH_2)_3{-}CH_3 \xrightarrow{H_2C:} CH_3{-}(CH_2)_4{-}CH_3 + CH_3{-}(CH_2)_2{-}CH(CH_3)_2 + CH_3CH_2CH(CH_3)CH_2CH_3$$

## 3.2 Special features of radical chain reactions

A radical, because of its unpaired electron, once formed will react as rapidly as possible with molecules around it or with another radical to satisfy the normal valency requirement, the driving force of the reaction being the energy liberated in bond formation. Thus to start up a radical reaction initiating radicals are introduced into the system containing the reactant or mixture of reactants, which then suffer attack by the initiating radicals to generate new radicals and a reaction chain is set up. This is referred to as the initiation step of the reaction and may be brought about in a variety of ways, one being through irradiation with ultraviolet light which produces the initiating radicals by homolysis and another being the addition of a reagent which easily suffers homolysis to supply such radicals. It should be noted here that only small quantities of the initiators are required, because once the reaction is started the new radicals generated can react further, i.e., propagate the reaction, by any of the processes of abstraction, addition, fragmentation, etc., which lead ultimately to the consumption of a large number of substrate molecules. This chain character, observed in the majority of homolytic reactions, is the most distinctive feature of radical chemistry. The chain may be terminated by any of the radical destroying processes such as coupling or disproportionation.

The three stages involved in chain reactions, namely, initiation, chain propagation and chain termination, are illustrated below in the chlorination of methane and the addition of bromotrichloromethane to ethylene.

*Radical chain chlorination of methane*

$$Cl_2 \xrightarrow{h\nu \text{ or } \Delta H} 2Cl\cdot \quad \text{initiation}$$

$$\left.\begin{aligned} Cl\cdot + CH_4 &\rightarrow CH_3\cdot + HCl \\ CH_3\cdot + Cl_2 &\rightarrow CH_3Cl + Cl\cdot \end{aligned}\right\} \text{chain propagation}$$

$$\left.\begin{aligned} Cl\cdot + Cl\cdot &\rightarrow Cl_2 \\ CH_3\cdot + Cl\cdot &\rightarrow CH_3Cl \\ CH_3\cdot + CH_3\cdot &\rightarrow C_2H_6 \end{aligned}\right\} \text{chain termination}$$

*Radical addition of bromotrichloromethane to ethylene*

$$BrCCl_3 \rightarrow Br\cdot + \cdot CCl_3 \quad \text{initiation}$$

$$\left.\begin{array}{r} Cl_3C\cdot + CH_2{=}CH_2 \rightarrow Cl_3CCH_2CH_2\cdot \\ Cl_3CCH_2CH_2\cdot + BrCCl_3 \rightarrow Cl_3CCH_2CH_2Br + \cdot CCl_3 \end{array}\right\} \text{chain propagation}$$

$$\left.\begin{array}{r} Cl_3C\cdot + Cl_3CCH_2CH_2\cdot \rightarrow Cl_3CCH_2CH_2CCl_3 \\ 2Cl_3C\cdot \rightarrow C_2Cl_6 \end{array}\right\} \text{chain termination}$$

One important feature of these chain processes to be pointed out is that at any one time the concentration of the propagating radical species is extremely low. This being so, chain terminating reactions become relatively unimportant and high conversion of the reactants can be realized through the chain propagating steps. On the other hand, small amounts of inhibitors which can trap or scavenge the propagating radicals are sufficient to slow down or inhibit the reaction altogether. Traces of such inhibitors are in fact directly responsible for the induction period in certain reactions.

Due to their high reactivity, the reaction of radicals with other molecules is usually rapid and of low activation energy. Consequently it is not surprising to find that quite often more than one reaction path is taken simultaneously, resulting in the formation of complex products. In a good many cases, however, one particular reaction path is favoured, and a large number of radical reactions have indeed proved useful in synthetic chemistry. Even in such cases side reactions almost inevitably occur to a certain extent with the consequence that near quantitative yields, often realized from heterolytic reactions, are seldom achieved.

Free radicals, being uncharged entities, are relatively insensitive to polar effects either in the solvent or at the reaction site. This contributes in part to the relatively low selectivity of their reactions. We shall see later that polar influences do in fact affect radical reactions, though not nearly to the same extent as they do ionic reactions. In general, radical stability is the most important governing factor, and the course that the reaction takes will be by way of the formation of the more stable radical. This in turn will depend largely on the energetics of the reaction and the strengths of the bonds involved. Thus, the reaction will be expected to proceed in such a way that the weakest bond is broken first, so that one finds that among the hydrogen atoms of alkanes, for instance, the tertiary hydrogen atom is the most readily abstracted, followed by secondary and then primary hydrogen atoms. One finds also that reactive radicals, like fluorine atoms, are relatively unselective in the type of hydrogen they abstract (because the reaction is extremely exothermic) while the less reactive radicals such as bromine atoms are more selective (the reaction here being endothermic).

## 3.3 The transition state

It is necessary to consider the energetics involved in a reaction to understand the factors governing it, such as the reactivity of the attacking radical, the relative rates of individual component steps of the reaction and their exothermicities. For the reaction between a radical R· and a substrate X—Y to occur an energy barrier would have to be passed to reach the transition state $R \cdots \dot{X} \cdots Y$, as shown in Fig. 3.1. (We shall ignore here

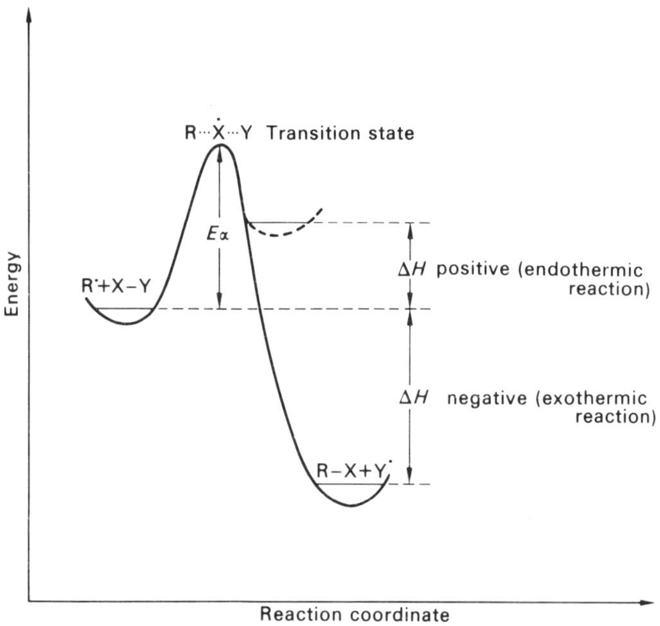


**Fig. 3.1** Activation energy, transition state and heat of reaction.

the possibility of tunnelling through the barrier which is quantum mechanically possible, especially for small atoms.) The height of this barrier, known as the activation energy $E_\alpha$, governs the rate of reaction and for most radical reactions is only about 15 kcal mole$^{-1}$ or less. This means that in the transition state there is no complete breakage of the X—Y bond, a process which requires energy of the order of 50–100 kcal mole$^{-1}$ for most compounds, but that both the attacking and the departing radicals are partially bonded to X. As is also shown, the reaction is exothermic if the heat of reaction $\Delta H$ is negative or endothermic if $\Delta H$ is positive, these being important considerations since the value of $\Delta H$ will determine whether the reaction will take place (that is, if the entropy term $\Delta S$ is not considerable). The energy of activation $E_\alpha$ can be equal to, but not smaller than, $\Delta H$ and it follows that strongly endothermic steps cannot play an important role in chain propagation.

The rate of a chemical reaction in relation to the activation energy and temperature is usually formulated by means of the Arrhenius equation

$$k = A\, e^{-E_\alpha/RT}$$

or by the equation from the transition state theory

$$k = \frac{\kappa T}{h} e^{\Delta S^\ddagger/R}\, e^{-\Delta H^\ddagger/RT}$$

where $\kappa$ is the Boltzmann constant, $h$ is Planck's constant, $T$ is the absolute temperature, $R$ is the gas constant, $\Delta S^\ddagger$ is the entropy of activation and $\Delta H^\ddagger$ is the heat of activation. The above equations are related by the following expressions

$$A = \frac{e\kappa T}{h} e^{\Delta S^\ddagger/R}$$

$$\Delta H^\ddagger = E_\alpha - RT \simeq E_\alpha \quad \text{(at room temperature)}$$

While it would be useful to be able to estimate the magnitudes of $E_\alpha$, as this would enable the rates of reactions to be determined and hence the nature of the products, this is in fact difficult to achieve. Although several rules based on bond dissociation energies have been devised only very limited success has so far been attained. An equally challenging goal would be a description, however approximate, of the transition state.[1] This has been depicted in Fig. 3.1 as having a centre atom placed symmetrically between the incoming and the leaving radicals. In practice, however, this is not at all likely to be so, except for the case where the attacking and departing radicals are identical; the extent of bond formation or bond breakage will vary according to the species involved. Furthermore, the transition state may also be profoundly affected by steric and polar factors.

A useful approach towards a description of the transition state in terms of its relative resemblance to the reactants or the products of a reaction has been formulated. According to Leffler's approximation[2] the transition state will resemble the least stable of the species among the reactants or products. Similarly, in terms of Hammond's postulate[3] the nature of the transition state will vary according to whether the reaction is exothermic or endothermic, being more reactant-like (A) in an exothermic reaction and more product-like (B) in an endothermic reaction, as illustrated in Fig. 3.2. Since (A) resembles the reactants the extent of bond breakage is likely to be small and the reaction would thus be insensitive to the nature of the bond to be

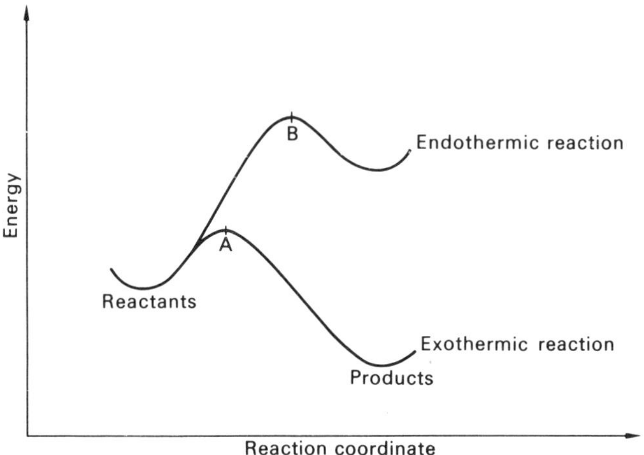


**Fig. 3.2** Transition state and exothermicity of reactions.

broken. Transition state (B) on the other hand being more product-like involves extensive bond breakage; the rate of reaction will therefore be affected by the nature of the bond being ruptured.

(A) *Bond dissociation energies*

To be able to predict, however approximately, the relative stabilities of radicals and the relative energies of radical reactions, data on the energy required to break specific bonds will be required. This quantity, known as the bond dissociation energy[4] is the heat of reaction on homolysis of the species concerned and for a species A—B is denoted by $D(\text{A—B})$. By this definition $D(\text{A—B})$ is also the activation energy $E_\alpha$ for the reaction

$$\text{A—B} \rightarrow \text{A}\cdot + \text{B}\cdot \qquad \Delta H = D(\text{A—B})$$

assuming that the reverse reaction, recombination, proceeds with little or no activation energy. The bond dissociation energy, it should be pointed out, is specific for a particular bond and is different from what is frequently referred to as bond strength, which is an average quantity, as illustrated in the case of methane:

$$CH_4 \rightarrow CH_3\cdot + H\cdot \qquad \Delta H = 104 \text{ kcal mole}^{-1}$$

$$CH_4 \rightarrow C + 4H\cdot \qquad \Delta H = 392 \text{ kcal mole}^{-1}$$

The bond dissociation energy of $CH_3$—H, denoted as $D(CH_3\text{—H})$, is 104 kcal mole$^{-1}$ while the average bond energy of methane is 392/4 or 98 kcal mole$^{-1}$. Since bond dissociation energy is more pertinent to radical chemistry future references to bond strength or bond energy will, unless stated otherwise, refer only to this quantity.

There are a number of methods available for the determination of bond dissociation energies. These will not be discussed here, but it should be noted that different methods may not give exactly the same values for a particular compound and unless corrections are made inconsistencies may appear in different texts. From the data on bond dissociation energies given in Tables 3.1–3.3 one important observation which may be made is that the bond dissociation energy is dependent on the structures of the radical fragments resulting from dissociation, and hence yields information on the relative stabilities of the radicals produced. Some care, however,

**Table 3.1** Selected Bond Dissociation Energies*

| Species | *D* | Species | *D* | Species | *D* |
|---|---|---|---|---|---|
| H—H | 104 | HO—H | 119 | Me—Me | 88 |
| Me—H | 104 | $CH_3CO_2$—H | 112 | Et—Me | 85 |
| Et—H | 98 | $CH_3OO$—H | 102 | $Pr^i$—Me | 84 |
| Pr—H | 98 | $O_2NO$—H | 102 | $Bu^t$—Me | 80 |
| $Pr^i$—H | 95 | $HO_2$—H | 90 | $Bu^t$—$Bu^t$ | 68 |
| $Bu^t$—H | 91 | PhO—H | 85 | $PhCH_2$—Me | 72 |
| $PhCH_2$—H | 85 | ONO—H | 79 | $CH_2$=$CHCH_2$—Me | 72 |
| $CH_2$=$CHCH_2$—H | 85 | ClO—H | 78 | MeC(=O)—Me | 82 |
| $Ph_3C$—H | 75 | F—H | 136 | MeC(=O)—C(=O)Me | 70 |
| $C_6H_7$—H† | 74 | Cl—H | 103 | $HOCH_2$—Me | 83 |
| HC≡C—H | 125 | Br—H | 88 | $C_{19}H_{15}$—$CPh_3$‡ | 15 |
| Ph—H | 112 | I—H | 71 | Ph—Ph | 100 |
| $CH_2$=CH—H | 103 | $H_2N$—H | 103 | Ph—Me | 93 |
| $CH_3C(=O)$—H | 88 | HS—H | 90 | Me—F | 108 |
| HOC(=O)—H | 90 | Cl—Cl | 58 | Me—Cl | 84 |
| $HOCH_2$—H | 93 | Br—Br | 46 | Me—I | 56 |
| $F_3C$—H | 104 | I—I | 36 | Me—OH | 92 |
| $Cl_3C$—H | 96 | F—F | 38 | Me—$NH_2$ | 79 |
| | | MeS—SMe | 73 | | |

* Data in kcal $mole^{-1}$, from ref. 4.

† $C_6H_7$ = (cyclohexa-2,5-dienyl structure)

‡ $C_{19}H_{15}$ = $Ph_2C$=(cyclohexadienyl ring bearing H)

**Table 3.2** Bond Dissociation Energies of some Radical Sources[4]

| Species | $D$ (kcal mole$^{-1}$) |
|---|---|
| HO—OH | 51 |
| $H_2N—NH_2$ | 58 |
| HO—Cl | 60 |
| MeN=N—Me | 46 |
| ONO—Me | 56 |
| MeHg—Me | 52 |
| $Cl_3C—Br$ | 49 |
| $Bu^tO—OBu^t$ | 37 |

**Table 3.3** Bond Dissociation Energies of some Radicals[4]

| Species | $D$ (kcal mole$^{-1}$) |
|---|---|
| $\cdot CH_2—Me$ | 96 |
| $\cdot CH_2CH_2—H$ | 39 |
| $\cdot CH_2CH_2—Me$ | 26 |
| $\cdot OCH_2—Me$ | 12 |
| $\cdot OCMe_2—Me$ | 7 |
| $O=\dot{C}—Me$ | 11 |

must be exercised as comparisons can only be made when the bonds being broken are of a very similar type, i.e., having similar extent of overlap, hyperconjugation and delocalization to other atoms. Clearly, this cannot be achieved in practice but for a particular series of closely related compounds it may be taken to be approximately true. For example, it is clear that for a series of radicals derived from alkanes the order of stability decreases in the order: tertiary, secondary, primary, and free methyl. This order may be explained in terms of stabilization due to hyperconjugation and/or the lower energy, due to release of steric strain in radical formation, in the more highly substituted radicals.

A comparison of the bond dissociation energies of propane (Pr—H) and propene ($CH_2=CHCH_2—H$) shows a difference of 13 kcal mole$^{-1}$. This is ascribed to the resonance energy in the allyl radical formed from propene.

The resonance energy of the free benzyl radical is of the same magnitude. By comparing triphenylmethane ($Ph_3C—H$) and isobutane ($Me_3C—H$), the resonance energy of triphenylmethyl is obtained as 16 kcal $mole^{-1}$ which, as has been noted from e.s.r. data, is not significantly greater than that of free benzyl. This is attributed to the non-planarity in the radical brought about by steric repulsion between the *ortho* hydrogen atoms. The exceptionally low value for the dimer of triphenylmethyl is also noteworthy and is due to steric and resonance effects.

The compounds listed in Table 3.2 have relatively low bond dissociation energies and are useful as radical sources. The ease of breakage of bonds $\beta$ to radical centres may also be noted from the values quoted in Table 3.3 and accounts for the widespread occurrence of fragmentation reactions. Particularly striking among them is the low value for the t-butoxy radical, which breaks down readily to free methyl and acetone.

Bond dissociation energy data also affords valuable information on heats of reaction and hence enables an assessment to be made of the relative ease with which reactions will occur. For instance, the reaction of any radical R· with HI would be expected to lead, almost certainly, to RH and I· rather than RI and H·, since the endothermicity of the latter pathway is forbiddingly large. In the abstraction of aliphatic hydrogen atoms by atomic bromine, one notes that the reaction is endothermic for both the secondary and the primary hydrogen atom, although in terms of bond dissociation energy the former is favoured by 3 kcal $mole^{-1}$. It follows that the activation energies, which must be in excess of the endothermicities of the reactions, must also differ by about 3 kcal $mole^{-1}$, and this if translated to relative rates by the equation $k = A\,e^{-E\alpha/RT}$ will give rates in favour of the secondary hydrogen atom by a factor of about 200 at room temperature. It can thus be seen that small differences in bond energies can lead to large differences in reaction rates.

From bond dissociation energy data fluorination and chlorination, but not bromination, of alkanes are seen to be highly exothermic reactions. These, indeed, occur with great avidity (see Chapter 11).

## (B) *Kinetic deuterium isotope effects*

In a reaction in which a carbon–hydrogen bond is broken, theory as well as experiment has shown that the rate of reaction will be reduced when the hydrogen atom is replaced by deuterium. This is referred to as a primary kinetic deuterium isotope effect, and its magnitude has frequently served as a guide to the structure of the transition state.[5–8]

Before considering the origin of the isotope effect it would be instructive to scrutinize the difference in characteristics of the C—H and C—D bonds,

**Table 3.4** Characteristics of C—H and C—D Bonds

| Property | Difference | Remarks on C—D bond |
|---|---|---|
| Mass | C—D > C—H | |
| Strength | C—D > C—H | Vibrational energy is inversely proportional to the square root of mass, hence lower zero-point energy (see Fig. 3.3) |
| Length | C—H > C—D | Lower mean square amplitude of vibration and lower zero-point energy in an asymmetric potential well means shorter bond length (see Fig. 3.3) |
| Steric size | C—H > C—D | Difference arising from shorter bond length |
| Electron density | C—D > C—H | do. |
| Electron donation | C—D > C—H | do. |
| Hyperconjugation | C—H > C—D | do. |
| Polarizability | C—H > C—D | do. |
| Tunnelling | C—H > C—D | Quantum mechanical tunnelling through activation barrier is less important for larger mass. |

which arise primarily from the difference in mass of the hydrogen and deuterium atoms. A summary of these bond characteristics is given in Table 3.4.

Deuterium isotope effects can broadly be classified into two types, namely, primary kinetic isotope effect and secondary isotope effects. The first type, of magnitude about 2–8, arises from the breaking of bonds to hydrogen or deuterium atoms and originates from differences in C—H and C—D bond strengths. Secondary isotope effects are smaller, with $k^{H}/k^{D}$ of around 1.5, and do not involve the breakage of bonds to hydrogen or deuterium atoms but only their effects on the reaction site. Secondary deuterium isotope effects are usually attributed to differences in steric or hyperconjugative factors between the two isotopes. Examples of deuterium isotope effects in radical reactions will be discussed later in this Chapter (Section 3.5).

To explain the origin of the primary deuterium isotope effect a simple model can be given. As shown in Fig. 3.3 the zero-point energy of the C—D

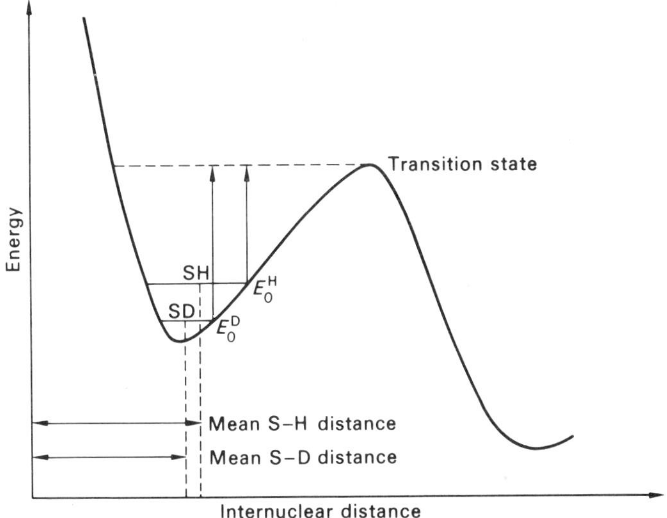


**Fig. 3.3** Potential energies of molecules HS and DS and their transition states.

bond, $E_0^D$, is lower than that of the C—H bond, $E_0^H$. As can be seen the loss of zero-point stretching vibration energy in the transition state gives rise to the primary kinetic isotope effect. This zero-point energy difference is about 1.15 kcal mole$^{-1}$ at room temperature and thus the isotope effect $k^H/k^D$ calculated would be about 7. This magnitude varies and depends on the amounts of bond forming and bond breaking or the symmetry of the transition state. (Westheimer[6] has in fact presented theoretical arguments to show that for a series of related hydrogen transfers the kinetic isotope effect $k^H/k^D$ will be a maximum when the transition state is symmetrical.) Primary kinetic isotope effects of 2–8 are generally encountered, larger values being ascribed to quantum mechanical tunnelling. Thus, the magnitude of the kinetic isotope effect which also frequently shows a correlation with the activation energy and other parameters will give an idea of the symmetry of the transition state and the extent of bond breaking.

## (C) *The Hammett equation*

The best known empirical relationship for correlating structure and reactivity is the Hammett equation.[9] For the reaction of a series of *meta*- and *para*-substituted benzene derivatives the relationship may be written as

$$\log (k/k_0) = \rho\sigma$$

where $k$ and $k_0$ are the rate or equilibrium constants for the reaction of the substituted and unsubstituted compound respectively. The reaction constant $\rho$ is a measure of the sensitivity of the reaction to changes in electron density at the reaction site, while the substituent constant $\sigma$ is a measure of the extent to which the substituent alters the electron density at the reaction site and is independent of the nature of the reaction involved. The value of $\sigma$ is defined from the dissociation constants of *m*- and *p*-substituted benzoic acids $X—C_6H_4—COOH$ at 25 °C, as follows,

$$X—C_6H_4—COOH + H_2O \rightleftarrows X—C_6H_4—COO^- + H_3O^+$$

$$\log (k/k_0) = \sigma$$

where $k$ is the dissociation constant for ionization of a substituted benzoic acid and $k_0$ that of the unsubstituted benzoic acid.

Other linear free energy relationships have now been devised, one of the most useful being that of Brown and Okamoto,[10] as follows:

$$\log (k/k_0) = \rho\sigma^+$$

The new substituent constants $\sigma^+$ are based on the solvolysis of the *t*-cumyl chlorides $X—C_6H_4—CMe_2Cl$ in 90% acetone–water, involving the following transition state:

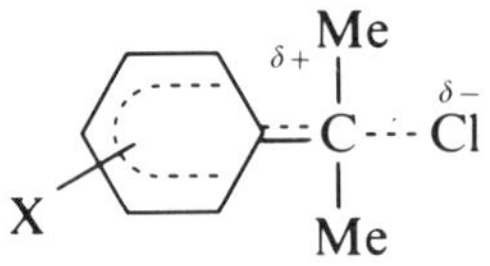

$$\text{X—C}_6\text{H}_4\overset{\delta+}{=\!=}\text{C}(\text{Me})_2\cdots\overset{\delta-}{\text{Cl}}$$

A nuclear substituent can exert its polar influence in the transition state principally in two ways, *via* inductive effects (through $\sigma$-bonds) and/or resonance effects (through $\pi$-bonds). From the definition of $\sigma^+$ constants it is clear that Brown and Okamoto's equation is intended for strongly polar transition states where resonance stabilization by the *p*-substituent plays an important role. In general, reactions in which such polar transition states are involved can be well correlated with $\sigma^+$ constants.

It is obvious that linear free energy relationships are useful in the study of reaction mechanisms and in some cases form the basis on which prediction can be made as to the course reactions would take. To date, no less than several thousand Hammett and Hammett-like plots have been worked out for a large variety of chemical reactions. Among these are an increasing number of radical reactions as more and more of them are found to be sensitive to polar effects. For the majority of these the Hammett relationships have been found to provide satisfactory correlation in terms of either the $\sigma$ or $\sigma^+$ constants, even though these relationships were originally defined for ionic reactions. As will be discussed later, for a number of these radical reactions correlations have actually been found to be better with $\sigma^+$ than with $\sigma$, showing that the transition states may well be highly polarized.

## 3.4 Structure and reactivity of radicals

The structure of the transition state, whatever its position along the reaction coordinate, will depend on the nature of the bond partially formed and the bond partially broken. When one considers the breakage of carbon–hydrogen bonds by radical abstraction one finds that the activation energies for such reactions increase in the order tertiary, secondary, and primary, which is also the order of increasing bond dissociation energies. In terms of the reactivity of the hydrogen atom abstracted one arrives at the decreasing order tertiary, secondary, and primary which in fact has been found to hold for abstraction by all radicals.

It has long been recognized that in general radicals which are very reactive, e.g. fluorine and chlorine atoms, show little discrimination in their attack on the different types of bonds. Less reactive radicals such as atomic bromine or trichloromethyl, on the other hand, show greater selectivity in their reactions. The reasonable assumption has thus been made that a relationship exists between the reactivity of a radical species and its selectivity, namely that high reactivity is associated with low selectivity, and *vice versa*.

The selectivity of a radical is partly dependent on the energy of the new bond formed; the greater the dissociation energy of this bond the less selective is the attack by the radical.[11,12] This is illustrated by the data shown in Table 3.5. Among the halogens the reactivity is greatest in fluorine, then chlorine followed by bromine, and this is in the same order as the bond dissociation energies in the respective hydrogen halides. The phenyl radical is similar in reactivity to the methyl radical; it is more selective than chlorine and t-butoxy radicals, but less selective than bromine and trichloromethyl radicals.

It is instructive to compare the reactivities of various carbon–hydrogen bonds towards bromine and chlorine radicals as given in Table 3.6. Here we find that the difference in the selectivity of bromine and chlorine atoms is very striking, as witnessed by the fact that, while towards bromine the $\alpha$-hydrogen in cumene is some two million times more reactive than the hydrogen atoms in ethane, towards chlorine the relative reactivity is only seven times. It is clear in the case of abstraction by bromine that the stabilization effects of the substituents on the radical being formed are in operation. This is understandable in terms of Hammond's postulate that, where there is considerable bond rupture the transition state (1a) would resemble the product (the radical

**Table 3.5** Relative Reactivities (per active hydrogen) of Alkanes and Arylalkanes towards Radicals*

| X· (°C) | Alkanes $-CH_3$ | $>CH_2$ | $>CH$ | $E_\alpha$† kcal mole$^{-1}$ | $D$(H—X) kcal mole$^{-1}$ |
|---|---|---|---|---|---|
| F· (25°) | 1 | 1.2 | 1.4 | 1.21 | 136 |
| HO· (17.5°) | 1 | 4.7 | 9.8 | | 119 |
| Cl· (25°) | 1 | 4.6 | 8.9 | 3.83 | 103 |
| MeO· (230°) | 1 | 8 | 27 | | 102 |
| $F_3C$· (182°) | 1 | 6 | 36 | 11.7 | 104 |
| Bu$^t$O· (40°) | 1 | 10 | 44 | | 106 |
| Ph· (60°) | 1 | 9.3 | 44 | 7.5 | 112 |
| Me· (182°) | 1 | 7 | 50 | 14.5 | 104 |
| $Cl_3C$· (190°) | 1 | 80 | 2300 | | 90 |
| Br· (98°) | 1 | 250 | 6300 | 18.25 | 87 |

| X· (°C) | Arylalkanes $PhCH_3$ | $PhCH_2Me$ | $PhCHMe_2$ | $Ph_2CH_2$ | $Ph_3CH$ | $D$(H—X) |
|---|---|---|---|---|---|---|
| Cl· (40°) | 1 | 2.5 | 5.5 | 2.0 | 7.2 | 102 |
| Bu$^t$O· (40°) | 1 | 3.2 | 6.8 | 4.7 | 9.6 | 106 |
| Ph· (60°) | 1 | 4.6 | 9.7 | 7.5 | 39 | 112 |
| Me· (65°) | 1 | 4 | 13 | — | — | 104 |
| ROO· (90°) | 1 | 7.8 | 13 | 16 | 13 | 90 |
| Br· (40°) | 1 | 17 | 37 | 10 | 18 | 87 |
| $Cl_3C$· (40°) | 1 | 50 | 260 | 50 | 160 | 90 |

* Data from A. F. Trotman-Dickenson, *Adv. Free-Radical Chem.*, 1965, **1**, chap. 1, Logos, London.

† For hydrogen abstractions from methane.

**Table 3.6** Effect of Structure on the Relative Reactivities of C—H Bonds towards some Radicals[13]

| Bond | $D(\gt C—H)$ | Br· (40 °C) | Cl· (40 °C) | Ph· (60 °C) | Bu$^t$O· (40 °C) |
|---|---|---|---|---|---|
| Me—H | 104 | 0.0007 | 0.004 | — | — |
| Et—H | 98 | (1) | (1) | (1) | (1) |
| Pr—H | 95 | 220 | 4.3 | 9.2 | 10 |
| Bu$^t$—H | 91 | 19 400 | 6.0 | 40 | 44 |
| $PhCH_2$—H | 85 | 64 000 | 1.3 | 9.2 | 10 |
| PhCH(Me)—H | — | 1 000 000 | 3.3 | 37 | 32 |
| $PhC(Me)_2$—H | — | 2 330 000 | 7.3 | 81 | 68 |
| $Ph_2CH$—H | — | 620 000 | 2.6 | 62 | 47 |
| $Ph_2C(Me)$—H | — | 2 700 000 | — | — | — |
| $Ph_3C$—H | 75 | 1 140 000 | 9.5 | 330 | 96 |

intermediate) and hence resonance and hyperconjugation stabilization effects play a dominant role. The transition state in chlorination (1b), on the other hand, involves little breaking of the carbon–hydrogen bond and therefore resembles the reactants and in turn is relatively insensitive to stabilization effects.

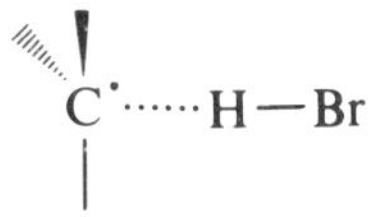

C$^{\bullet}$······H—Br (1a)          C—H······Ċl (1b)

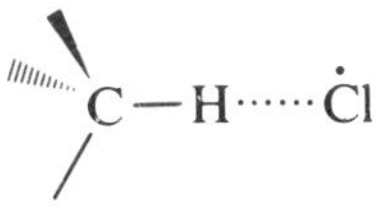

Another aspect of the relationship between structure and reactivity may be discerned from the data in Table 3.6. Successive replacement by methyl (or phenyl) groups in the substrate does not produce as large an activation of the hydrogen to be abstracted as the first replacement. For instance, in abstraction by bromine from propane, ethylbenzene and diphenylmethane the reactivity first increases but then drops slightly; similarly from toluene, ethylbenzene and cumene the first reactivity increment is seen to be much greater than the second. This 'saturation effect'[13] is similar to that observed in bond energy and radical stabilization as noted in the previous section. An obvious conclusion to be drawn from these observations is that there are factors influencing radical reactions, other than resonance stabilization effects, which must be taken into account. These factors, which may have polar or steric origins, will be discussed in Chapter 9.

As to the selectivity and the energy of activation ($E_\alpha$) in hydrogen abstraction from alkanes the data from Table 3.5 show that these appear closely related and vary in a fairly regular manner. In the case of radicals which form bonds of comparable strengths to hydrogen such as chlorine, methyl and trifluoromethyl, a comparison of their selectivities can be made. If we assume that the polar effect does not operate in these hydrogen abstraction reactions, it is seen that of the radicals mentioned chlorine is the least selective.

It is clear from the above discussion that the stability or reactivity of a radical has an important bearing on the way it reacts. In Section 3.3 it was stated that the relative stability of radicals can be approximately predicted from thermodynamic considerations of bond strengths. However, there are limitations to this method. Firstly, as pointed out earlier, only the strengths of closely similar types of bonds can be used for evaluation of the relative stability of the corresponding radicals. Secondly, it is often difficult to obtain accurate bond dissociation energy data.

Other ways of comparing radical stability have been devised. For example, the ease of fragmentation, relative to other reaction paths, of a radical can be taken as an indication of the stability of the new radical R· so formed. Thus the extent to which each of the radicals $Ph\dot{C}HOR$ (R = Ph, Me, Et, $Pr^i$, $Bu^t$, and $PhCH_2$) breaks down to R· and benzaldehyde, relative to the extent of dimerization, gives a measure of the relative stability of the radical R· (see also Chapter 4). Other entities which find similar use are the *t*-alkoxy radical $RCMe_2O\cdot$ and the dialkoxyalkyl radical $Me\dot{C}(OR)OBu^n$ (Chapter 14). These reactions are outlined below:

$$Ph\dot{C}HOR \longrightarrow (R\cdot + PhCHO) \textit{ versus } [PhCHOR]_2$$

$$RCMe_2O\cdot \longrightarrow (R\cdot + Me_2CO) \textit{ versus } (Me\cdot + RCOMe)$$

$$Me\dot{C}(OR)OBu^n \longrightarrow (R\cdot + MeCO_2Bu^n) \textit{ versus } (Bu^n\cdot + MeCO_2R)$$

In another method (Rüchardt)[14] the radical R· produced from the *t*-butyl perester $RCO_2OBu^t$ is allowed to react competitively with carbon tetrachloride and bromotrichloromethane and the relative rate constant $k_{Br}/k_{Cl}$ so obtained is taken as a measure of the reactivity of the radical (see scheme given). Polar effects are assumed to be approximately the same in both reactions since the leaving radical in each case is trichloromethyl.

$$R-\overset{\overset{\displaystyle O}{\|}}{C}-O-OBu^t \longrightarrow R\cdot + CO_2 + Bu^tO\cdot$$

$$R\cdot + CCl_4 \xrightarrow{k_{Cl}} RCl + \cdot CCl_3$$

$$R\cdot + BrCCl_3 \xrightarrow{k_{Br}} RBr + \cdot CCl_3$$

From the results shown in Table 3.7 certain inferences may be drawn concerning the factors which govern radical selectivity. The data for phenyl radicals shows that selectivity increases as the temperature is decreased whereas nuclear substituents do not cause any significant change. Among the cyclic radicals, the cyclopropyl radical, which is non-planar, exhibits relatively low selectivity. Noteworthy also is the unusually low selectivity of 'bridgehead' radicals which are constrained to be non-planar. However, it should be noted that, while the scale of radical selectivities given here affords a general indication of the relative stabilities of the radicals, it does not constitute an exact measure of these quantities. This is because other factors which influence the above reactions (see Chapters 4 and 9) have been assumed to be negligible.

**Table 3.7** Relative Reactivities of Radicals[14]

| R· | $T$ (°C) | $k_{Br}/k_{Cl}$ |
|---|---|---|
| Ph· | 130° | 144 |
| Ph· | 104° | 184 |
| Ph· | 80° | 257 |
| 4-MeO$C_6H_4$· | 130° | 129 |
| 4-Cl$C_6H_4$· | 130° | 124 |
| cyclo-$C_3H_5$· | 110° | 278 |
| cyclo-$C_4H_7$· | 110° | 573 |
| cyclo-$C_5H_7$· | 110° | 662 |
| cyclo-$C_6H_{11}$· | 110° | 566 |
| cyclo-$C_8H_{15}$· | 110° | 451 |
| Ph$CH_2$· | 80° | 1700 |
| 9-tripticyl· | 130° | 33 |
| 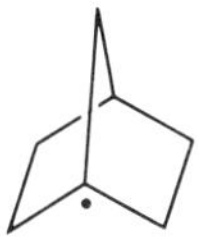  | 130° | 47 |
| 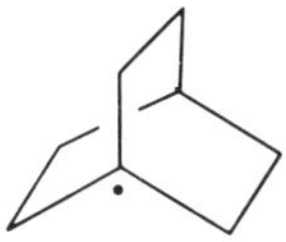  | 80° | 59 |
|   | 80° | 29 |
| 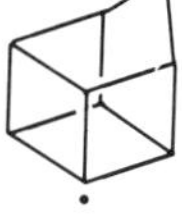  | 130° | 73 |

## 3.5 Kinetic deuterium isotope effects in radical reactions

Since the magnitude of the kinetic isotope effect is very much dependent on the structure of the transition state its application in the study of reaction mechanisms has been widespread. The size of the effect, as pointed out earlier, bears a parallel relationship to the activation energy and the Hammett $\rho$-value (i.e., polar effect), as for instance in the hydrogen abstraction by atomic bromine from toluene, ethylbenzene and cumene, for which the magnitudes of both the activation energy and the kinetic isotopic effect decrease in the same order (see Chapter 9). This is not unexpected since each parameter is dependent on the structure of the transition state.[15]

In general a small isotope effect is observed for a hydrogen abstraction process requiring small activation energy, as for instance in the abstraction of benzylic or tertiary hydrogen atoms by atomic chlorine. As mentioned earlier such reactions are characterized by high exothermicity and relatively little breakage of the C—H bond in the transition state because this occurs fairly early along the reaction coordinate. Thus abstraction of the benzylic hydrogen atom in toluene-$d_1$ and the tertiary hydrogen in isobutane gives $k^H/k^D$ values of the same magnitude, *ca.* 1.3, whereas less exothermic processes such as abstraction from ethane and dideuteromethane giye values of 2.68 and 12.1, respectively, corresponding to increasing bond breakage in the transition state in these latter cases. Similarly bromination of toluene is less exothermic than chlorination and thus gives a larger isotope effect. This

**Table 3.8** Primary Deuterium Isotope Effects in Hydrogen Abstraction Reactions[15]

| Substrate | Radical (source) | Solvent (T °C) | $k^H/k^D$ |
|---|---|---|---|
| $Me_3C—H$ | Cl· ($Cl_2$) | reactant (−15°) | 1.3–1.5 |
| Ethane | Cl· ($Cl_2$) | reactant (27°) | 2.68 |
| $CH_2D_2$ | Cl· ($Cl_2$) | reactant (0°) | 12.1 |
| Toluene | Cl· ($Cl_2$) | reactant (77°) | 2.1 |
| Toluene | Cl· ($Cl_2$) | $CCl_4$ (77°) | 1.3 |
| Toluene | Cl· ($Cl_2$) | reactant (110°) | 1.47 |
| Toluene | Cl·, ·$SO_2Cl$($SO_2Cl_2$) | reactant (110°) | 1.73 |
| Toluene | Cl· (NCS) | reactant (110°) | 1.59 |
| $PhCH_2D$ | Br· ($Br_2$) | reactant (77°) | 4.6 |
| $PhCHD_2$ | Br· ($Br_2$) | reactant (77°) | 4.9 |
| $PhCH_2D$ | Br· (NBS) | reactant (77°) | 4.9 |
| $PhCHD_2$ | Br· (NBS) | reactant (77°) | 5.0 |
| Toluene | Br· (NBS) | reactant (110°) | 3.59 |
| Toluene | $CH_3COO$· ($CH_3CO_2$)$_2$ | reactant (120°) | 9.9 |
| *p*-$ClC_6H_4CHD_2$ | Br· ($Br_2$) | reactant (77°) | 5.73 |
| *p*-$ClC_6H_4CHD_2$ | Br· (NBS) | reactant (77°) | 5.54 |
| *p*-$MeOC_6H_4CH_2D$ | Br· (NBS) | reactant (77°) | 3.22 |
| $PhCH_2CH_3$ | Cl· ($Cl_2$) | reactant (80°) | 1.4 |
| $PhCH_2CH_3$ | Br· (NBS) | reactant (77°) | 2.67 |
| $PhCH_2CH_3$ | $CH_3COO$· ($CH_3CO_2$)$_2$ | reactant (120°) | 6.5 |
| $PhCHMe_2$ | Br· (NBS) | reactant (77°) | 1.81 |
| Toluene | $Bu^tO$·, Cl· ($Bu^tOCl$) | reactant | 5.5 |
| Toluene | $Bu^tO$· ($Bu^tON_2OBu^t$) | reactant | 5.0 |
| $Ph_2CHD$ | $CH_3COO$· ($CH_3CO_2$)$_2$ | reactant (120°) | 6.7 |

and other data are listed in Table 3.8, from which yet another piece of information can be obtained. Thus atomic chlorine in carbon tetrachloride and in toluene is seen to give different isotope effects, and the larger value in the latter solvent might be taken as support for the postulation of a complexed chlorine atom. Pursuing this line of reasoning further, it should be possible to make use of data on the isotope effect to identify the abstracting species. This has actually been done. For example, the fact that the same value for $k^H/k^D$ was obtained for bromine and for *N*-bromosuccinimide (NBS) at the same temperature, while *N*-chlorosuccinimide (NCS) gave a completely different value, affords conclusive evidence that the first two reagents involve the same reacting species, that is, atomic bromine. The earlier postulate that NBS reacts in the form of the succinimidyl radical, which would have given a $k^H/k^D$ value identical with that for NCS, is thus invalid (see also Section 11.2).

The primary isotope effect discussed above involves hydrogen transfer in the rate determining step. Secondary deuterium isotope effects[7,16] are also possible for reactions not involving the rupture of hydrogen bonds in the transition state, but these effects are small, $k^H/k^D$ being of the order of 1.2. These may be divided into two types: the $\alpha$-effect, in which the deuterium atom is attached to the carbon atom undergoing a spatial (or hybridization) change during the reaction, and the $\beta$-effect in which the isotopic hydrogen is on a more remote carbon which does not undergo such a change. The former type is usually attributed to zero-point vibrational (bending mode) energy changes during the course of the reaction while the latter effect arises from hyperconjugation.

Data on secondary isotope effects in radical reactions are relatively sparse and only a few instances will be quoted here.[17] In the $\alpha$-effect, the substitution of deuterium on a carbon atom, which changes from $sp^3$ to $sp^2$ hybridization during reaction, has been estimated to be about 12% per deuterium, i.e., $k^H/k^D$ of *ca.* 1.1 per deuterium atom. If the carbon atom undergoes a hybridization change in the reverse direction, from $sp^2$ to $sp^3$, then an inverse isotope effect is observed, $k^H/k^D$ being then *ca.* 0.9. For example, in the homolysis of azo-*bis*-$\alpha$-phenylethane labelled in both $\alpha$-positions with deuterium, $k^H/k^D$ was found to be 1.27, corresponding to about 12% per deuterium if both bonds break simultaneously. The addition of the methyl radical to $CH_3CH{=}CD_2$ and $PhCD{=}CD_2$, and of polystyryl radical to $PhCH{=}CD_2$ gave for $k^H/k^D$ values of 0.89, 0.89, and 0.88, respectively, the terminal carbon atom changing hybridization from $sp^2$ to $sp^3$ in these cases. The $k^H/k^D$ values of 1.06 and 1.04 for the decomposition of $PhCD(Me)CO_2OBu^t$ and of $PhCH(CD_3)CO_2OBu^t$, are examples of $\alpha$- and $\beta$-isotope effects, respectively. Instances of the simultaneous operation of the two effects are also known, as in the addition of methylene to butene-2-$d_8$

and of hydrogen atoms to propylene-$d_6$, in both of which cases $k^H/k^D$ was 1.08.

Although secondary isotope effects are of theoretical interest, they are generally not of much value in providing information on the geometry of the carbon atoms in the transition state since (a) their magnitudes are small and (b) there is no general agreement as to which is the main contributing factor (loss of zero-point energy in rehybridization, steric, hyperconjugation or polar effects).

## 3.6 Kinetics of radical chain reactions

The study of the kinetics of radical chain reactions is attended by many complications as the rate expressions are generally complex and the reactions are often sensitive to the inadvertent presence of even traces of catalysts and inhibitors. Nevertheless a number of such studies have successfully been made of gas phase reactions, among which the classical example of the reaction between hydrogen and bromine vapour, $H_2 + Br_2 \rightarrow 2HBr$, serves as an illustration.[18] To obtain a kinetic expression for the reaction the complete reaction sequences including initiation, propagation and termination steps are first written down, as indicated:

$$Br_2 \xrightarrow{k_1} 2Br\cdot \qquad (1)$$

$$Br\cdot + H_2 \xrightarrow{k_2} HBr + H\cdot \qquad (2)$$

$$H\cdot + Br_2 \xrightarrow{k_3} HBr + Br\cdot \qquad (3)$$

$$H\cdot + HBr \xrightarrow{k_4} H_2 + Br\cdot \qquad (4)$$

$$2Br\cdot \xrightarrow{k_5} Br_2 \qquad (5)$$

A few points about this scheme should be noted. Firstly, bromine and hydrogen atoms, the chain carriers, are present only in very low concentrations. Secondly, reaction (2), being strongly endothermic, is slow while reactions (3) and (4) are relatively rapid and it follows therefore that the concentration of the more stable bromine atoms is always higher than that of atomic hydrogen.

The rates of the formation of HBr, Br·, and H· may be expressed by the following terms:

$$\frac{d[HBr]}{dt} = k_2[Br\cdot][H_2] + k_3[H\cdot][Br_2] - k_4[H\cdot][HBr] \qquad (6)$$

$$\frac{d[Br\cdot]}{dt} = 2k_1[Br_2] - k_2[Br\cdot][H_2] + k_3[H\cdot][HBr_2] - 2k_5[Br\cdot]^2 + k_4[H\cdot][HBr] \qquad (7)$$

$$\frac{d[\mathrm{H}\cdot]}{dt} = k_2[\mathrm{Br}\cdot][\mathrm{H}_2] - k_3[\mathrm{H}\cdot][\mathrm{Br}_2] - k_4[\mathrm{H}\cdot][\mathrm{HBr}] \tag{8}$$

A rigorous solution is not possible but an approximate one can be achieved by applying the steady state approximation. This states that the concentration of the chain carrier, which is very low at all times, remains essentially constant. In other words, if one were to measure the concentration of the intermediate radical, say by e.s.r., one would find this increasing initially and then reaching a constant value. Equations (7) and (8) can then be equated to zero, and from equations (6) and (8) one gets

$$\frac{d[\mathrm{HBr}]}{dt} = 2k_3[\mathrm{H}\cdot][\mathrm{Br}_2] \tag{9}$$

and from (7) and (8),

$$2k_1[\mathrm{Br}_2] - 2k_5[\mathrm{Br}\cdot]^2 = 0$$

or

$$[\mathrm{Br}\cdot] = \frac{k_1}{k_5}[\mathrm{Br}_2]^{\frac{1}{2}} \tag{10}$$

Substitution of equation (10) into (8) gives

$$[\mathrm{H}\cdot] = \frac{k_2\left(\dfrac{k_1}{k_5}\right)^{\frac{1}{2}}[\mathrm{H}_2][\mathrm{Br}_2]^{\frac{1}{2}}}{k_3[\mathrm{Br}_2] + k_4[\mathrm{HBr}]}$$

Using this expression, equation (9) may then be solved to give:

$$\frac{d[\mathrm{HBr}]}{dt} = \frac{2k_2\left(\dfrac{k_1}{k_5}\right)^{\frac{1}{2}}[\mathrm{H}_2][\mathrm{Br}_2]^{\frac{1}{2}}}{1 + \dfrac{k_4}{k_3}\dfrac{[\mathrm{HBr}]}{[\mathrm{Br}_2]}}$$

which agrees with the expression obtained empirically, *viz*.,

$$\frac{d[\mathrm{HBr}]}{dt} = \frac{\text{constant}\,[\mathrm{H}_2][\mathrm{Br}_2]^{\frac{1}{2}}}{1 + \text{constant} \times \dfrac{[\mathrm{HBr}]}{[\mathrm{Br}_2]}}$$

Generally chain mechanisms can be satisfactorily handled by the steady state approximation to give suitable kinetic expressions and these may be tested experimentally. A brief treatment of the kinetics of addition and polymerization will be given later in Chapter 12; for details of this subject the reader is referred to standard texts such as Walling's.[19]

### 3.7 Conservation of orbital symmetry in organic reactions

Since its recognition a few years ago, the principle of the conservation of orbital symmetry[20] has found wide application not only in the study of reaction mechanism, but also in predicting new reactions and assigning characteristics to them. The results of the application of this principle led to the famous Woodward-Hoffman rules which govern a large number of reactions of both ground state as well as excited state molecules, of which the latter (Photochemistry) are directly connected with radical chemistry. Additionally these rules afford an explanation as to why some concerted reactions which are 'forbidden' due to orbital symmetry restrictions proceed via radical pathways.

The details of the subject of orbital symmetry conservation will not be dealt with here as there are many excellent reviews[20] available, but the case of cycloaddition reactions will be mentioned briefly. The well-known Diels-Alder reaction of an olefin and a diene may be referred to as a [2 + 4]-cycloaddition while the cyclization of two olefins to give cyclobutane may be called a [2 + 2]-cycloaddition. The latter reaction may be imagined as proceeding via a concerted transition state, as follows:

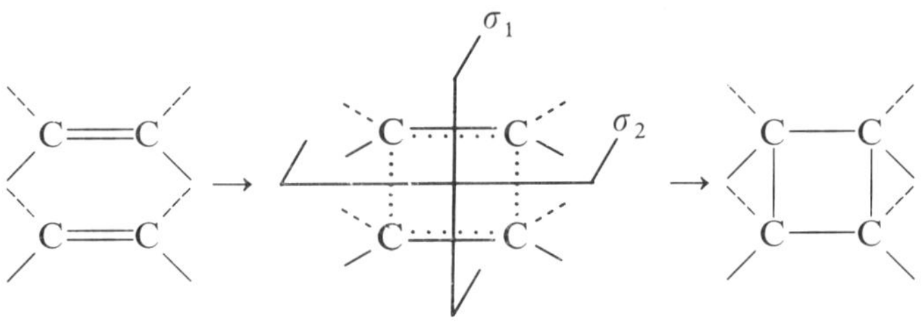

According to Woodward and Hoffman one could qualitatively account, from a consideration of the symmetries of the molecular orbitals involved in the reaction (in the present case with respect to the $\sigma_1$ and $\sigma_2$ planes of symmetry), for the ease or difficulty of the cycloaddition. For a concerted reaction to occur, i.e., symmetry allowed, the bonding ground state orbitals of the reactant must be correlated with the bonding orbitals of the product. Likewise the antibonding (excited state) orbitals must be similarly correlated.[21] A [2 + 2]-cycloaddition turns out to be symmetry forbidden. Thermal reactions of this type, as predicted, are rare and when they do occur they follow non-concerted mechanisms. On the other hand, a reaction involving one excited state (e.g., a photochemical process) is predicted to be symmetry allowed and therefore should be facile. [2 + 4]-cycloadditions may similarly be predicted to be symmetry allowed, and this explains the widespread occurrence of thermally induced [2 + 4]-cycloadditions but not that of the [2 + 2] type. On the other hand, photochemical [2 + 2]-cycloadditions, but not the [2 + 4] variety, are very common, and this is again in accord with theory. Symmetrical carbene additions to olefins (Chapter 17) which may be considered as [1 + 2]-cycloadditions, are also predicted to be ground state forbidden reactions. But since carbene additions are facile and widespread, the reactions are not 'forbidden' but are 'allowed' because the transition state of the reaction is likely to be unsymmetrical.

Several interesting aspects of [2 + 2]-cycloadditions have been experimentally found to be in accordance with theoretical predictions. For example concerted dimerization of *cis*-butene may be realized photochemically but not thermally. Similarly the thermal decomposition of the 1,2-dioxetan (2) to ground state ketone, which being the reverse reaction of [2 + 2]-cycloaddition is also a 'forbidden' reaction. This means that (2)

C—O
|   |
C—O

(2)

Cl   F
 C=C
Cl   F

(3)

in the ground state cannot decompose directly to give ketones in their ground states. In agreement with theory, it has been found that one of the product ketones is in fact in its excited state (see Section 7.3). This reaction is the basis of most of the chemi- and bio-luminescent reactions now understood. In another instance, it was found that 1,1-dichloro-2,2-difluoroethylene (3) dimerizes readily to a cyclobutane derivative.[22] Again since a concerted reaction of this type is forbidden on orbital symmetry grounds, the reaction proceeds via diradical intermediates as has been shown to be the case (see Section 12.9).

## References


1 M. Szwarc in *The Transition State*, The Chemical Society, Special Publication No. 16, London, 1962, p. 91.
2 J. E. Leffler, *Science N.Y.*, 1953, **117**, 340.
3 G. S. Hammond, *J. Am. chem. Soc.*, 1955, **77**, 334.
4 S. W. Benson, *J. chem. Educ.*, 1965, **42**, 502.
5 F. G. Bordwell and W. J. Boyle, Jr., *J. Am. chem. Soc.*, 1971, **93**, 514.
6 F. H. Westheimer, *Chem. Rev.*, 1961, **61**, 265.
7 L. Melander, *Isotope Effects on Reaction Rates*, Ronald Press, New York, 1960.
8 H. Simon and D. Palm, *Angew. Chem., Int. Ed.*, 1966, **5**, 920.
9 L. P. Hammett, *Physical Organic Chemistry*, McGraw-Hill, New York, 1970, p. 355.

10 H. C. BROWN and Y. OKAMOTO, *J. Am. chem. Soc.*, 1958, **80**, 4979.

11 A. F. TROTMAN-DICKENSON, *Adv. Free-Radical Chem.*, 1965, **1**, chap. 1, G. H. WILLIAMS ed., Logos, London.

12 J. M. TEDDER, *Q. Rev. chem. Soc.*, 1960, **14**, 336.

13*a* G. A. RUSSELL and C. DEBOER, *J. Am. chem. Soc.*, 1963, **85**, 3136;

*b* R. F. BRIDGER and G. A. RUSSELL, *ibid.*, 1963, **85**, 3754.

14 C. RÜCHARDT, *Angew. Chem., Int. Ed.*, 1970, **9**, 830.

15*a* K. B. WIBERG, *Chem. Rev.*, 1955, **55**, 713;

*b* K. B. WIBERG, *J. Am. chem. Soc.*, 1958, **80**, 3033.

16 E. A. HALEVI *Prog. Phys. Org. Chem.*, 1964, **1**, 109, S. G. COHEN, A. STREITWIESER, JR., and R. W. TAFT eds., Interscience, New York.

17*a* T. W. KOENIG and W. D. BREWER, *Tetrahedron Lett.*, 1965, 2773;

*b* T. KOENIG, J. HUNTINGTON, and R. CRUTHOFT, *J. Am. chem. Soc.*, 1970, **92**, 5413;

*c* W. A. PRYOR, R. W. HENDERSON, R. A. PATSIGA, and N. CARROLL, *J. Am. chem. Soc.*, 1966, **88**, 1199.

18 A. A. FROST and R. G. PEARSON, *Kinetics and Mechanism*, J. Wiley, New York, 1962, p. 236.

19 C. WALLING, *Free Radicals in Solution*, J. Wiley, New York, 1957, Chapter 3.

20 R. B. WOODWARD and R. HOFFMANN, *a Angew. Chem., Int. Ed.*, 1969, **8**, 11; *b Acc. Chem. Res.*, 1968, **1**, 17; *c The Conservation of Orbital Symmetry*, Academic Press, 1970, New York.

21 E. M. KOSOWER, *An Introduction to Physical Organic Chemistry*, J. Wiley, New York, 1968, p. 207.

22*a* L. K. MONTGOMERY, K. SCHUELLER, and P. D. BARTLETT, *J. Am. chem. Soc.*, 1964, **86**, 622;

*b* P. D. BARTLETT, G. E. H. WALLBILLICH, A. S. WINGROVE, J. S. SWENTON, L. K. MONTGOMERY, and B. D. KRAMER, *J. Am. chem. Soc.*, 1968, **90**, 2049.

# 4
# Conformation of Alkyl Radicals

In an alkyl radical three possibilities may be envisaged for the preferred geometry of the radical centre. The three bonds attached to the centre may be in the same plane, or they may be directed to the corners of a regular tetrahedron or those of a shallow pyramid, as illustrated below:

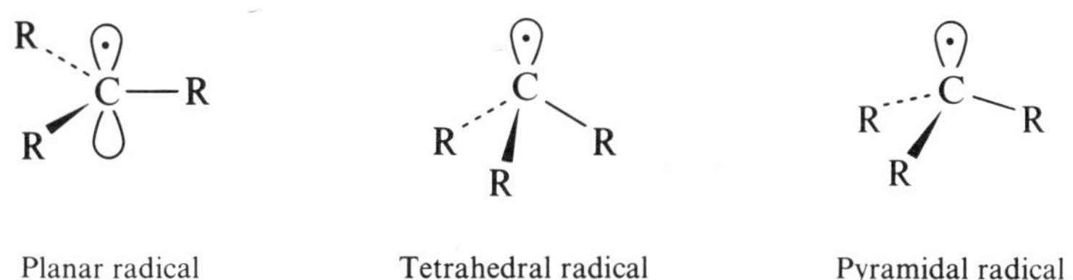

Planar radical | Tetrahedral radical | Pyramidal radical

The bulk of evidence obtained by ultraviolet, infrared, Raman, and e.s.r. spectroscopic methods, carried out on the methyl radical in particular, was consistent with either a planar structure or a shallow, rapidly inverting pyramidal structure, but was inadequate to differentiate between the two. Results from theoretical calculations were similarly inconclusive in this respect, although they showed that the energy difference between the two conformations is very small, contrary to the case between the corresponding conformations in carbanions and carbonium ions. Recently, however, electron spin resonance studies of the methyl and other alkyl radicals (Symons)[1] have yielded convincing evidence in favour of the planar conformation. In such a conformation the hybridization of the bonds would be $sp^2$, and the unpaired electron would find itself in the *p*-orbital.

During recent years another line of evidence has been made available through chemical studies of a number of fused ring and cage systems, from the rigid structures of which a number of radicals with unequivocally pyramidal conformations have been generated and compared with radicals which exist or are capable of existing in the planar structure. These pyramidal radicals include 1-adamantyl, 1-bicyclo[2.2.2]octyl, apocamphyl, *cis* and *trans* decalinyl, and tripticyl, shown below:

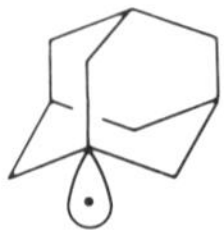

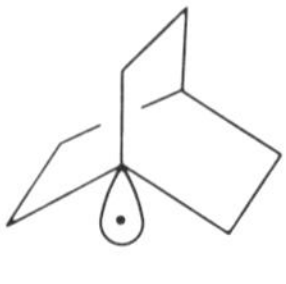

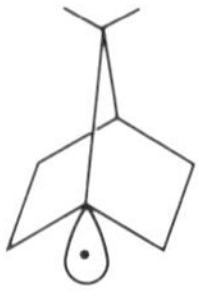

1-Adamantyl 1-Bicyclo[2.2.2]octyl Apocamphyl

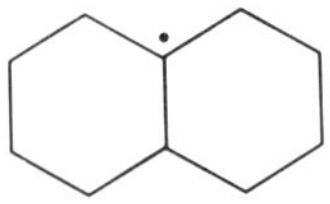

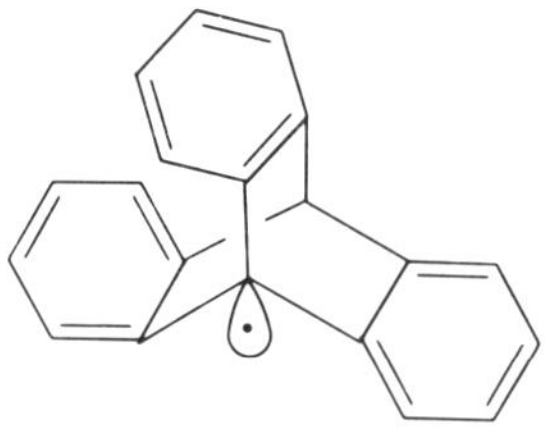

*cis/trans* Decalinyl Tripticyl

The pyramidal structure of two of these radicals, the 1-adamantyl and the bicyclo-octyl, has in fact been demonstrated by e.s.r. spectroscopy.[2]

For the thermal decomposition of the perester $RCO_2OBu^t$ three independent kinetic studies[3] have given the following scale of relative rates, for variations in R:

$$t\text{-Butyl} \geqslant \text{1-adamantyl} > \text{1-bicyclo[2.2.2]octyl} > \text{1-norbornyl}$$

If the perester decomposed by a concerted two-bond scission, as indicated,

$$RCO_2OBu^t \rightarrow R\cdot + CO_2 + Bu^tO\cdot$$

the relative rates at which the radicals R· are formed could be taken as reflecting their relative stabilities, and thus the inference could be drawn that the preferred geometry in these alkyl radicals is the planar one. This inference is, however, subject to reservations since it is now known that polar transition states are involved in the decomposition of the perester and consequently the rate would depend on the stability of the radical R· as well as that of the ion.

The decomposition of *cis* and *trans* *t*-butyl-9-decalinpercarboxylate ($RCO_2OBu^t$) in the presence of oxygen, which proceeds by the following mechanism to give a mixture of the *cis* and *trans* 9-decalinols (ROH), has been studied by Bartlett and co-workers.[4] Their results, listed in Table 4.1, show that, whereas for the *trans* perester the ratio of *trans* ROH to *cis* ROH obtained is independent of oxygen pressure, *trans* ROH being the main product, the same ratio for the *cis* perester is dependent on oxygen pressure and at the highest pressure employed *cis* ROH becomes the major product. These observations have been accounted for as follows (see Figure). Whereas the *trans* decalinyl radical (derived from the *trans* perester) requires only a slight flexing of its structure to achieve a planar conformation, for the *cis* radical this would involve inversion of one of its chair-form rings. This latter process may well be slow relative to the time taken to react with oxygen when this is present in a high concentration, hence resulting in *cis* ROH being the main product. With low concentrations of oxygen, however, the *cis* radical has time to assume the planar structure, and does so before reacting with oxygen, to give both *cis* and *trans* ROH. The fact

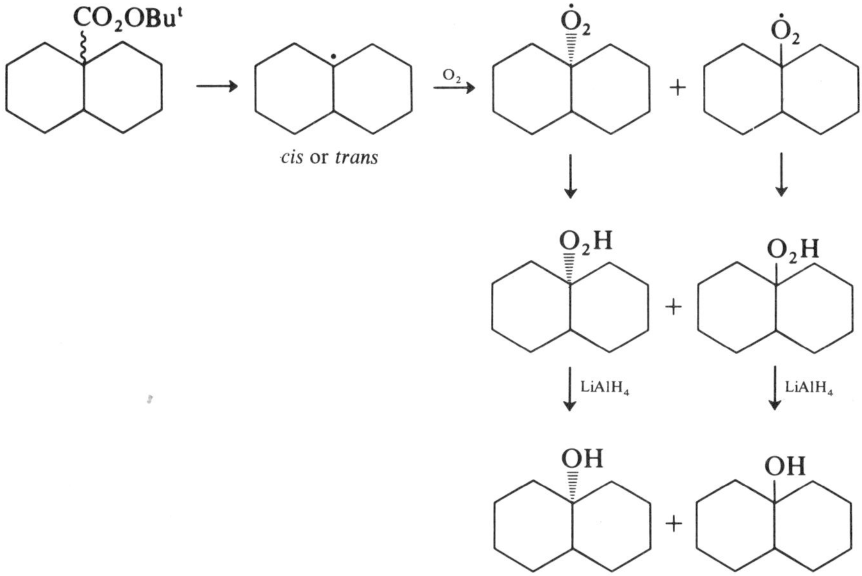

**Table 4.1** Decomposition of *cis* and *trans* *t*-butyl-9-decalinpercarboxylate

| Perester | $O_2$ (atmospheres) | *trans* ROH : *cis* ROH |
|---|---|---|
| *trans* | 1 | 7 |
| | 60 | 7 |
| *cis* | 1 | 7 |
| | 545 | 0.4 |

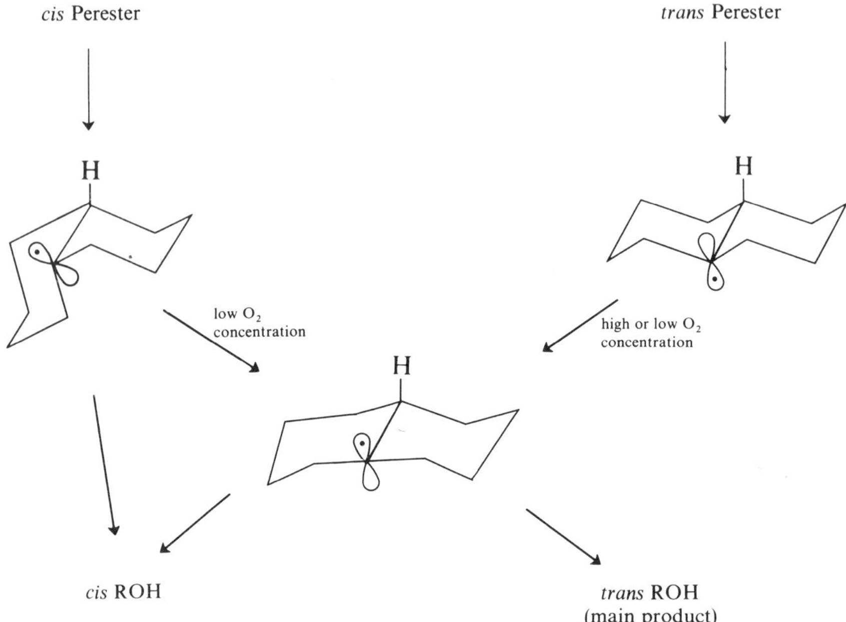


that the *cis* radical, which is in a pyramidal conformation, takes up the planar conformation if given time to do so is clear indication that the latter is the more stable.

The conformation of the 9-decalinyl radical has also been studied using *cis* and *trans* 9-decalinylcarbinyl hypochlorites (ROCl), which decompose as shown (Greene):[5]

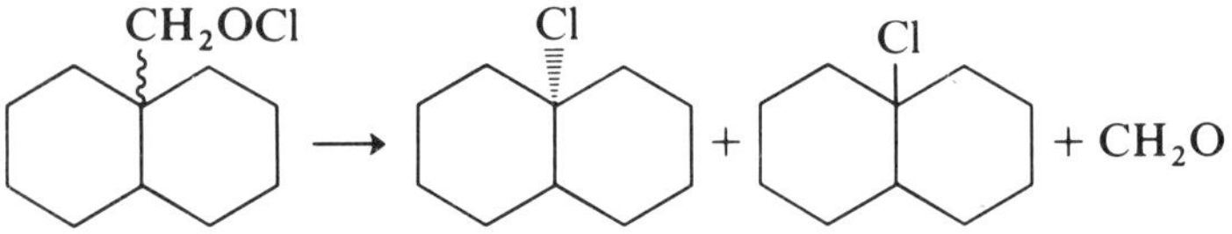

$CH_2OCl$ → Cl + Cl + $CH_2O$

The ratio of *trans* and *cis* RCl obtained from the *trans* hypochlorite is independent of initial hypochlorite concentration and favours the *trans* chloride (30:1 at −40 °C), while that from the *cis* hypochlorite is dependent on the initial concentration of hypochlorite, favouring the *cis* chloride at high concentration and the *trans* chloride at low concentration. These results may similarly be interpreted as evidence for the greater stability of the planar structure.

The tertiary alkoxy radical $RMe_2C—O\cdot$, produced from the hypochlorite $RMe_2C—OCl$ by thermolysis, may undergo (a) fragmentation to the radical

R· and acetone, R· then attacking the hypochlorite to give RCl, and/or (b) in the presence of cyclohexene abstract a hydrogen atom from it to liberate the 3-cyclohexenyl radical, which in turn reacts with the hypochlorite to give 3-cyclohexenyl chloride, as illustrated below:

$$RMe_2C{-}OCl \longrightarrow RMe_2C{-}O\cdot$$

$$RMe_2C{-}O\cdot \longrightarrow R\cdot + Me_2CO, \qquad R\cdot \longrightarrow RCl$$

$$RMe_2C{-}O\cdot \xrightarrow{\text{cyclohexene}} RMe_2C{-}OH + \text{(3-cyclohexenyl radical)} \longrightarrow \text{(3-chlorocyclohexene, Cl)}$$

Since the ease in which fragmentation takes place depends on the stability of the radical R· so produced, the ratio of RCl to the cyclohexenyl chloride affords a measure of the relative stability of R· when a series of the hypochlorites are decomposed under identical conditions. In this manner Greene et al.,[6] arrived at a scale of relative stabilities for R· as follows: isopropyl > ethyl > 1-norbornyl ~ methyl, showing once again the relative instability of the bridgehead pyramidal radical (see also Section 14.1).

Similarly, the relative stabilities of the apocamphyl and the tripticyl[7] radicals, as inferred from their relative ease of formation from the peroxides $(RCO)_2O_2$, have been equated to that of the methyl radical.

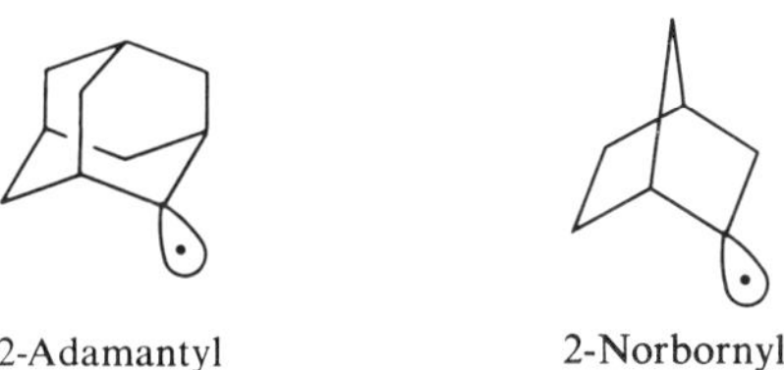

2-Adamantyl 2-Norbornyl

The $\alpha$-alkoxybenzyl radical $Ph\dot{C}HOR$, derived from the benzyl ether $PhCH_2OR$ through abstraction of a benzylic hydrogen atom by a radical such as the *t*-butoxy radical, had earlier been found[8] to undergo either or both of two processes, namely, (a) dimerization and (b) fragmentation to the radical R· and benzaldehyde, as follows:

$$PhCH_2OR \rightarrow Ph\dot{C}HOR \begin{cases} \nearrow [PhCHOR]_2 \\ \searrow R\cdot + PhCHO \end{cases}$$

$$R = PhCH_2, Ph_2CH, Bu^t, Pr^i, Et, Me, Ph$$

When fragmentation gives a relatively stable radical, such as benzyl, this process predominates, but when it leads to an energized radical, such as methyl or phenyl, its rate slows down and dimerization becomes the main reaction. The ratio of dimer to benzaldehyde *r* may thus be used as a measure of the relative stability of the radical R·, and in this way the following scale of relative stabilities of alkyl radicals has been obtained:

$$Ph_2CH, PhCH > Pr^i > Et > Me, Ph$$

Chick and Ong[9] have recently made use of the benzyl ether system to study the relative stabilities of a number of bridgehead radicals. From the dimer-benzaldehyde ratios *r* listed in Table 4.2, it is evident that the radicals with rigid pyramidal structures, namely, 1-adamantyl and 1-apocamphyl, are both of relatively low stability, comparable to that of methyl. Within the adamantyl system, it is also noted that the tertiary 1-adamantyl radical ($r = 2.5$) is considerably less stable than the secondary 2-adamantyl radical ($r = 0$), presumably because the latter is capable of existing in either a pyramidal ($sp^3$) or in a planar ($sp^2$) conformation. The same situation is found in the two bicycloheptyl radicals, 1-apocamphyl and 2-norbornyl.

**Table 4.2** Dimer-Benzaldehyde Ratio *r* from $Ph\dot{C}HOR$

| R: | Me | Et | Pr$^i$ | Bu$^t$ | $PhCH_2$ | $Ph_2CH$ |
|---|---|---|---|---|---|---|
| *r*: | 4.5 | 2.2 | 1.6 | 0.38 | 0 | 0 |

| R: | 1-adamantyl | 2-adamantyl | 1-apocamphyl | 2-norbornyl |
|---|---|---|---|---|
| *r*: | 2.5 | 0 | 3.6 | 0.07 |

The fact that the decomposition of 1,1′-azoadamantane requires an activation energy 20 kcal greater than does 2,2′-azoisobutane[10] furnishes further evidence for the relatively lower stability of the pyramidal structure.

## References


1 M. C. R. Symons, *Nature Lond.*, 1969, **222**, 1123.
2 P. J. Krusic, T. A. Rettig, and P. V. R. Schleyer, *J. Am. chem. Soc.*, 1972, **94**, 995.
3 J. P. Lorand, S. D. Chodroff, and R. W. Wallace, *J. Am. chem. Soc.*, 1968, **90**, 5266; R. C. Fort and R. E. Franklin, *J. Am. chem. Soc.*, 1968, **90**, 5267; L. B. Humphrey, B. Hodgson, and R. E. Pincock, *Can. J. Chem.*, 1968, **46**, 3099.
4 P. D. Bartlett, R. E. Pincock, J. H. Rolston, W. G. Schindel, and L. A. Singer, *J. Am. chem. Soc.*, 1965, **87**, 2590.
5 F. D. Greene and N. N. Lowry, *J. org. Chem.*, 1967, **32**, 875.
6 F. D. Greene, M. L. Savitz, F. D. Osterholtz, H. H. Lau, W. N. Smith, and P. M. Zanet, *J. org. Chem.*, 1963, **28**, 55.
7 S. F. Nelson and E. F. Travacedo, *J. org. Chem.*, 1969, **34**, 3651.
8 R. L. Huang and S. S. Si-Hoe, *Vistas in Free Radical Chemistry*, W. A. Waters ed., Pergamon Press, London, 1959, p. 242.
9 For a preliminary communication see W. H. Chick and S. H. Ong, *Chem. Commun.*, 1969, 216.
10 M. Prochazka, O. Ryba, and D. Lim, *Colln. Czech. chem. Commun.*, 1968, **33**, 3387.

# 5

# Production of Radicals

In this chapter a survey is given of the general methods available for the production of radicals, involving irradiation, redox reactions and thermal homolysis. Reference is also made to the more commonly used radical sources and their efficiency as initiators in homolytic reactions. A more detailed account of the three processes of radical generation mentioned, among which that of thermal homolysis has found the widest application, will appear in subsequent chapters.

## 5.1 Irradiation

As pointed out in Section 3.3 most bonds encountered in organic compounds have energies of the order of 50–95 kcal mole$^{-1}$. To bring about the cleavage of such bonds photolytically, light energy of this magnitude will be required. Table 5.1 shows that this can be achieved by use of visible or ultraviolet radiation; the latter, in fact, is of sufficient energy to effect the breakage of most types of bonds, provided the substance can absorb the radiation. Even when this proviso is not met, another substance called a

**Table 5.1** Energy Conversion Table*

| Å | nm | cm$^{-1}$ | kcal mole$^{-1}$ | eV |
|---|---|---|---|---|
| 2000 | 200 | 50 000 | 143.0 | 6.20 |
| 2500 | 250 | 40 000 | 114.4 | 4.96 |
| 3000 | 300 | 33 333 | 95.3 | 4.13 |
| 3500 | 350 | 28 571 | 81.7 | 3.54 |
| 4000 | 400 | 25 000 | 71.5 | 3.10 |
| 4500 | 450 | 22 222 | 63.5 | 2.76 |
| 5000 | 500 | 20 000 | 57.2 | 2.48 |

* From N. J. Turro *Molecular Photochemistry*, Benjamin, New York, 1967, p. 5.

'photosensitizer' can be added to promote energy transfer. It may be noted at this point that, if the energetics are favourable for the promotion and sustenance of a radical chain reaction, only small doses of the initiator (in this case light quanta) are necessary. For instance, in the reaction between hydrogen and chlorine, a single quantum of light can set off a chain reaction consuming $10^4$–$10^6$ molecules of the reactants.

Two advantages of photo-generation of radicals should be mentioned: (i) the reaction may be carried out at any convenient temperature and this makes low temperature studies possible, and (ii) the rate of radical production can be controlled by adjustment of the light intensity as well as the concentration of the absorbing species. On the other hand, the disadvantage of the reaction products being light sensitive must also be noted. Furthermore, use of ultraviolet radiation often requires reactions to be carried out in silica and not ordinary pyrex glass vessels.

Highly ionizing radiations (X-rays, $\gamma$-rays and cosmic particles) can undoubtedly also cause bond cleavage. In this case the energies involved are very far in excess of that required to rupture chemical bonds, so that not only are bonds cleaved, and cleaved indiscriminately, but atomic nuclei are also affected. Although this method may be of service in biological and certain other studies, it is of little importance where the controlled production of radicals of known structure is required. Generation of radicals by ultrasonics and various mechanical means are likewise of limited use and will not be discussed.

## 5.2 Redox reactions

Transition metal ions are useful redox agents. They decompose peroxides to generate radicals at relatively low temperatures. The combination of ferrous ion and hydrogen peroxide, a reagent discovered by Fenton in 1894, was the first such system to be used in organic chemistry. Redox initiators are useful for production of radicals in aqueous systems, and are particularly valuable as catalysts for low temperature emulsion polymerization.

Radical formation by electron transfer is not confined to systems involving metals and their ions. In fact, there is a large group of reactions which proceed by transfer of electrons between compounds, from electron-rich (or donors) to electron-deficient compounds (or acceptors).[1] The reaction, for example, of vinyl methyl ether with tetracyanoethylene in tetrahydrofuran gives first a red-orange charge-transfer complex (1) followed by electron transfer and collapse of the resulting diradical ion pair (2) to give a cyclobutane derivative. Such reactions are usually very sensitive to the polarity of the solvent. Again, the reaction of the carbanion of 2-nitropropane with 4-nitrocumyl chloride has been depicted by Kornblum[2] as proceeding by electron transfer to give first the radical ion (3) which breaks down to the radical (4), and this in turn combines with the nitropropyl radical to give the product (5).

MeO NC CN NC CN MeO NC CN NC CN (1)

MeŌ NC CN NC CN (2) MeŌ NC CN CN CN MeO CN CN CN CN

$$\mathrm{Me_2\bar{C}NO_2} + \mathrm{O_2N{-}C_6H_4{-}CMe_2Cl} \longrightarrow \mathrm{Me_2\dot{C}NO_2} + \mathrm{O_2N{-}C_6H_4{-}C\dot{\bar{M}}e_2Cl}\ (3)$$

$$(3) \longrightarrow \mathrm{O_2N{-}C_6H_4{-}\dot{C}Me_2}\ (4) + \mathrm{Cl^-}$$

$$(4) + \mathrm{Me_2\dot{C}NO_2} \longrightarrow \mathrm{O_2N{-}C_6H_4{-}CMe_2CMe_2NO_2}\ (5)$$

## 5.3 Thermal homolysis

The production of radicals for use in the temperature range in which reactions can most conveniently be carried out, about 50–150 °C, requires initiators with bond strengths of the order of 25–35 kcal mole$^{-1}$. This value can be calculated from the Arrhenius equation $k = A\,e^{-E_\alpha/RT}\ \mathrm{s}^{-1}$ for unimolecular bond homolysis, where $A$ may be taken to be $10^{-3}$–$10^{-4}$. Several groups of compounds which meet the above requirement, notably those containing the weak peroxo linkage, are available. These are listed in Table 5.2. They include a number of peroxides, peroxy esters and *bis* azo compounds and represent the most important thermal initiators available

**Table 5.2** Decomposition of Peroxides, Peroxy compounds and Azo compounds

| Compound | Approximate Half-lives in Hours* (°C) | $E_\alpha$ (kcal mole$^{-1}$) |
|---|---|---|
| MeO—OMe | $6 \times 10^4$ (60°) | 37 |
| EtO—OEt | $7 \times 10^3$ (60°) | 32 |
| Bu$^t$O—OBu$^t$ | $10^5$ (60°); $2 \times 10^2$ (100°); 2 (140°) | 37 |
| Bu$^t$O—OC(=O)—OBu$^t$ | $10^3$ (60°) | — |
| Bu$^t$O—OC(=O)—C(=O)O—OBu$^t$ | $10^{-2}$ (60°) | — |
| $MeCO_2$—OBu$^t$ | $10^4$ (60°); 10 (102°) | 38 |
| $PhCO_2$—OBu$^t$ | $5 \times 10^2$ (60°); 10 (105°) | 34 |
| $PhCH_2CO_2$—OBu$^t$ | 30 (60°) | 29 |
| $Ph_2CHCO_2$—OBu$^t$ | 0.4 (60°) | 24 |
| Bu$^t$O—OH | 5 (100°) | — |
| $PhCMe_2O$—OH | 25 (113°) | — |
| MeC(=O)O—OC(=O)Me | 1 (85°) | 31 |
| PhC(=O)O—OC(=O)Ph | 2 (90°) | 30 |
| $^-O_3SO—OSO_3^-$ | $2 \times 10^2$ (50°); 2 (80°) | 34 |
| Me—N=N—Me | $10^{13}$ (60°); 0.3 (300°) | 52 |
| $Me_2C(CN)$—N=N—$C(CN)Me_2$ | 17 (60°); 1 (85°) | 31 |
| Ph—N=N—$CHPh_2$ | 2 (54°) | 34 |
| Bu$^t$O—N=N—OBu$^t$ | 2 (55°) | — |
| Ph—N=N—$CPh_3$ | 1 (53°) | 27 |

* The values given are for comparative purposes and are only approximate. These will vary depending on the solvent and the extent of induced decomposition. For further details see C. Walling, *Free Radicals in Solution*, J. Wiley, New York, 1957, Chapter 10.

at present. A more detailed description of the decomposition of some of these compounds will be given in the next chapter.

The usefulness of each initiator is governed by a number of factors: (a) the temperature required for a suitable rate of decomposition, (b) the efficiency of radical production, and (c) the reactivity of the radicals generated.

The ready decomposition of azo compounds and peresters can be attributed to the formation of very stable products ($N_2$ and $CO_2$ respectively). The relationship between structure and the ease of decomposition may be explained by reference to Fig. 5.1, from which it is clear that increasing stability of R·, the radical being formed, decreases the activation energy for the decomposition. Thus, for the series of azo compounds R—N=N—R a high temperature is required for dissociation when R· are primary alkyl groups whereas decomposition becomes much more facile if R· is benzhydryl or triphenylmethyl, which are resonance stabilized. Strongly stabilized radicals so generated may however be too unreactive to be of use for initiating radical chains.

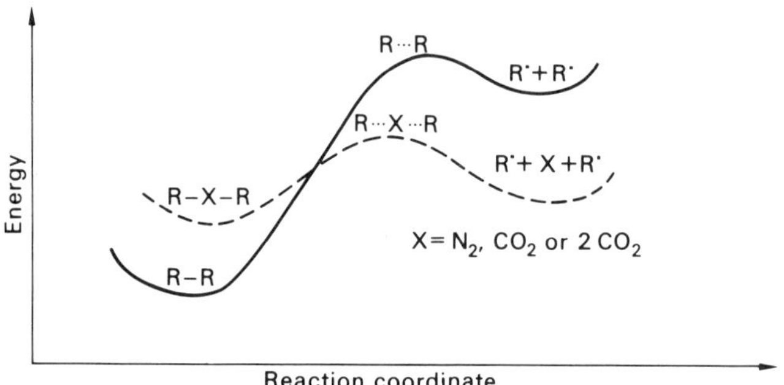


**Fig. 5.1** Homolysis of R–R and R–X–R.

A few comments will be made here on some of the more extensively used radical initiators. Azo *bis* isobutyronitrile $Me_2C(CN){-}N{=}N{-}CMe_2CN$ (often abbreviated to AIBN) is the best known of the azo type initiators, and is a particularly versatile reagent because (a) the temperature of decomposition (65–85 °C) is suitable for most reactions, (b) its first order rate of decomposition varies little with different solvents, (c) it is not susceptible to radical attack so that induced decomposition and transfer reactions are negligible, and (d) it may also be photolytically decomposed at relatively low temperatures.

Except for *t*-butyl peroxide, the other peroxides and peroxy compounds are in general dangerously explosive. Provided proper precautions are observed, benzoyl peroxide can be handled safely in the solid form but the others are used in the form of dilute solutions. For reactions making use of *t*-butoxy radicals choice of reaction temperature over a wide range (50–150 °C) is possible by employing *t*-butyl hyponitrite, *t*-butyl perbenzoate or *t*-butyl peroxide as the radical source. When a still lower reaction temperature is required irradiation can be applied.

## 5.4 Other radical-forming reactions

The formation of diradicals has been mentioned in the case of [2 + 2]-cycloadditions (Section 3.5). Rearrangements of carbanions involving 1,2 shifts also proceed by radical pair formation. These instances of radical production are non-chain in character.

The interaction of some non-radical species with each other can sometimes lead to rapid radical formation and the process is referred to as molecule-induced homolysis. Most instances known involve the interaction of unsaturated compounds with initiators. For example, styrene reacts spontaneously and exothermically with *t*-butyl hypochlorite at 0 °C, a reaction the radical nature of which was established by Walling[3] from the following observations: (a) oxygen or *t*-butyl catechol inhibits the reaction, (b) an *anti*-Markownikov addition product PhCHCl—$CH_2OBu^t$ is formed, (c) added cyclohexane is chlorinated although *t*-butyl hypochlorite and cyclohexane by themselves do not react at 0 °C in the dark, and (d) the reaction products are very similar to those of a light initiated process.

Such molecule-induced decomposition of initiators may be explained as due to the lowering of the activation energy of homolysis as a result of solvent participation in the transition state. Two possibilities of solvent interaction may be distinguished: one in which the solvent merely complexes with the incipient radicals and the other in which bond formation with the solvent molecule occurs. This may be illustrated in Fig. 5.2 where the initiator and solvent molecule are represented by X—Y and S, respectively. In the case of actual bond formation to give $\dot{S}$—X, the reaction can be exothermic if the bond dissociation energy *D*(S—X) is greater than *D*(X—Y). This is indeed possible in the reaction of styrene with *t*-butyl hypochlorite since for

$$Bu^tOCl + PhCH{=}CH_2 \rightarrow Bu^tO\cdot + Ph\dot{C}HCH_2Cl$$

$\Delta H = -9$ kcal mole$^{-1}$.

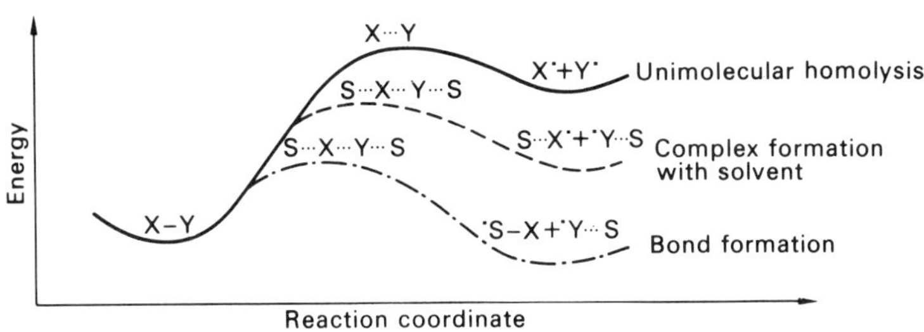


**Fig. 5.2** Molecule induced homolysis of initiator X–Y.

## 5.5 Initiator efficiency

The amount of an initiator to be used in a particular reaction depends on the initiating radical and the nature of the system involved. As the only function of an initiator is to produce radicals to start the reaction it would be desirable to be able to use as little initiator as possible. This condition is realized in many radical chain reactions, which in general are exothermic and for which consequently small quantities of initiator suffice. Only in cases where the chains are relatively short need greater amounts of initiator be employed, although this may lead to excessive incorporation of initiator fragments in the products. Just as chain lengths are governed by the energetics of the reaction, so the course initiation takes, whether this be by an abstraction or an addition process, is dependent on the strength of the bonds involved. Thus in the selection of an initiator the reactivity of the radical to which it gives rise has to be taken into consideration. The choice also depends on the temperature at which the reaction has to be carried out since the temperature range over which controllable decomposition occurs varies with different initiators.

Apart from thermodynamic and kinetic considerations, it has to be borne in mind that initiators have different capacities for producing radicals which will take an effective part in the desired reaction, i.e., initiating chains. Efficiencies in this respect in general fall short of 100% and this is because (a) the initiator may decompose in part by a mechanism which does not produce radicals, e.g., by an ionic mechanism, and/or (b) some of the radicals generated may not initiate radical chains but instead undergo other unproductive, or 'wastage,' reactions such as inducing the decomposition of initiator molecules, coupling or disproportionating with another radical, or participating in the termination of radical chains. An example of loss of efficiency due to a non-radical decomposition of the initiator is the Criegee rearrangement of per-esters (6). Radical induced decomposition of benzoyl peroxide, as outlined below and discussed in greater detail in Chapter 6, is also a 'wastage' reaction.

$$\mathrm{R'C(=O){-}O{-}O{-}CR_3} \longrightarrow \mathrm{R'C(=O){-}\overset{\delta-}{O}\cdots\overset{\delta+}{O}(R)\cdots CR_2} \longrightarrow \mathrm{R'C(=O){-}O{-}CR_2(OR)}$$

(6)

$$\mathrm{PhC(=O)O{-}OC(=O)Ph} \xrightarrow{\mathrm{Ph\cdot}} \mathrm{PhC(=O)O\cdot} + \mathrm{PhOC(=O)Ph}$$

The question of initiator efficiency is frequently an important one in the treatment of the kinetics of radical chain processes, e.g., polymerization.[4] If this efficiency, representing the fraction of initiator radicals which are successful in the initiation of radical chains, is denoted by $f$, the rate of formation of such chains by an initiator will be $2fk_d$[Initiator] where $k_d$ is the rate at which the initiator decomposes to give initiating radicals, as indicated:

$$\text{Initiator} \xrightarrow{k_d} 2\,\text{In}\cdot$$

The initiator would thus be 100% efficient if $f = 1$. In practice this is never attained and an initiator is considered quite efficient if $f$ is greater than 0.5. The value will vary appreciably with the initiator, the substrate, and the medium used.

While the extent to which most wastage reactions occur can be reduced by suitable dilution or by the right choice of reactive substrates, this cannot be done with cage reactions. The phenomenon of cage reactions and their effect on initiator efficiency may be illustrated by reference to the AIBN· system, which has been well investigated.[5] As mentioned earlier, this compound is not susceptible to induced decomposition and the rate of unimolecular homolysis remains fairly constant in a number of solvents. In the following scheme depicting the reactions of AIBN, two types of cyanopropyl radicals are shown, one remaining within the solvent cage (in square brackets) while the other, having broken free from the cage, is at liberty to react with surrounding molecules such as the solvent, an added scavenger, or a like radical.

$$Me_2\overset{CN}{\overset{|}{C}}{-}N{=}N{-}\overset{CN}{\overset{|}{C}}Me_2 \rightarrow \left[2Me_2\overset{CN}{\overset{|}{C}}\cdot + N_2\right]_{\text{cage}}$$

$$\left[2Me_2\overset{CN}{\overset{|}{C}}\cdot + N_2\right] \xrightarrow{\text{cage}} Me_2\overset{CN}{\overset{|}{C}}{-}N{=}C{=}CMe_2 \leftarrow 2Me_2\overset{CN}{\overset{|}{C}}\cdot \xrightarrow{\text{Scavenger}} \text{Products}$$

$$\left[2Me_2\overset{CN}{\overset{|}{C}}\cdot + N_2\right] \rightarrow 2Me_2\overset{CN}{\overset{|}{C}}\cdot$$

$$Me_2\overset{CN}{\overset{|}{C}}{-}N{=}C{=}CMe_2 \rightleftharpoons \left[2Me_2\overset{CN}{\overset{|}{C}}\cdot\right] \rightarrow 2Me_2\overset{CN}{\overset{|}{C}}\cdot$$

$$\left[2Me_2\overset{CN}{\overset{|}{C}}\cdot\right] \xrightarrow{\text{cage}} Me_2\overset{CN}{\overset{|}{C}}{-}\overset{CN}{\overset{|}{C}}Me_2$$

$$\left[2Me_2\overset{CN}{\overset{|}{C}}\cdot + N_2\right] \rightarrow Me_2\overset{CN}{\overset{|}{C}}{-}\overset{CN}{\overset{|}{C}}Me_2 \leftarrow 2Me_2\overset{CN}{\overset{|}{C}}\cdot$$

Both types of radicals can give two coupling products, a stable succinonitrile and an unstable ketenimine. Since the ketenimine can decompose (at about the same rate as AIBN) to cyanopropyl radicals again, it is only the succinonitrile that constitutes a wastage product. Radical scavengers may be expected to reduce the yield of succinonitrile derived from the 'free' radicals (but not that formed from the caged ones) to a limiting value, and this value therefore represents the amount of product formed within the solvent cage. Thus using butanethiol, a good hydrogen donor, as scavenger, the yield of succinonitrile has been found to decrease with increasing concentration of the scavenger until it finally levels off at about 20%. This means that 20% of the initiator radicals are wasted in cage recombination to give succinonitrile, or in other words that AIBN is 80% efficient in radical production under these circumstances. The efficiency of AIBN is in fact lower than this for most other systems since few substrates approach the reactivity of butanethiol as hydrogen donors.

The example given above illustrates the measurement of radical efficiency by analysis of the products. If the scavenger used is the stable diphenylpicrylhydrazyl (DPPH) which has an intense violet colour, the amount of cyanopropyl radicals can actually be estimated from the fading of DPPH colour. This method depends, of course, on the assumption of a quantitative reaction between DPPH and the radical.

$$Me_2\dot{C}CN + Ph_2N-\dot{N}-C_6H_2(NO_2)_3 \longrightarrow \text{non-radical products (light yellow colour)}$$

In the case of polymerization processes[4] initiator efficiencies can also be ascertained by estimation of initiator fragments incorporated in the polymer molecules and the measurement of the average molecular weight of the polymer. Since the concentrations of initiator fragments are usually very low, good use has been made of labelled initiators, such as carbon-14 labelled AIBN. In a hypothetical case of polymerization where every initiator yields two free radicals ($f = 1$) and if chains terminate by combination only, each initiator radical will produce only one molecule of polymer. If the number found is less than one, then the efficiency is accordingly lower. In practice it is necessary to know also the relative importance of radical combination and disproportionation in the termination process and to ensure that no induced decomposition has taken place. If induced decomposition has occurred (as with peroxides) a correction would have to be made.

Using AIBN labelled with carbon-14 efficiencies varying from 0.5 to almost 1.0 have been found for the polymerization of methyl methacrylate, vinyl acetate, styrene, vinyl chloride and acrylonitrile (increasing in this order).

## References


1 E. M. KOSOWER, *Prog. Phys. Org. Chem.*, 1965, **3**, 81

2 N. KORNBLUM, R. T. SWIGER, G. W. EARL, H. W. PINNICK, and F. W. STUCHAL, *J. Am. chem. Soc.*, 1970, **92**, 5513.

3 C. WALLING, L. HEATON, and D. D. TANNER, *J. Am. chem. Soc.*, 1965, **87**, 1715.

4*a* J. C. BEVINGTON, *Radical Polymerization*, Academic Press, London, 1961, p. 29.

*b* C. WALLING, *Free Radicals in Solution*, J. Wiley, New York, 1957, p. 67.

5 G. S. HAMMOND, C. S. WU, O. D. TRAPP, J. WARKENTIN, and R. T. KEYS, *J. Am. chem. Soc.*, 1960, **82**, 5394.

# 6
# Thermal Homolysis

At temperatures of 400–500 °C thermal excitation of molecules becomes strong enough to disrupt even the relatively stable carbon–carbon bonds in alkanes (bond energy of the order of 80 kcal mole$^{-1}$). The thermal cracking of petroleum is carried out in this temperature range. The radicals so produced containing more than two carbon atoms are unstable under these conditions and break down further, so that the total reaction is a complicated one involving a multiplicity of radicals and reaction pathways. Thus in the mechanism study of thermolysis the compounds which have yielded relatively precise information are understandably those containing comparatively weak bonds, fission of which to yield radicals occurs at lower temperatures than for the alkanes and similar thermally stable substances. Three groups of such compounds have been particularly well investigated. These are (a) the peroxides RO—OR′, R, and R′ being hydrogen, alkyl or acyl groups, (b) the azo compounds R—N═N—R′, R, and R′ being alkyl groups, and (c) organometallic compounds of such metals as lead, mercury, and magnesium. In these substances the bonds which undergo thermolysis readily, in the convenient temperature range of 40–150 °C, are the peroxidic, the azo, and the metal–carbon linkage respectively (bond energies of the order of 25–35 kcal mole$^{-1}$). Many of these compounds are extensively used radical sources for the initiation of homolytic reactions.

## 6.1 Peroxides

Thermolysis of peroxides in general proceeds by scission of the peroxidic bond as follows:

$$\text{RO—OR}' \rightarrow \text{RO}\cdot + \cdot\text{OR}'$$

However, with certain peroxides a bimolecular process, known as induced decomposition, occurs as a side reaction arising from attack by another radical, usually one derived from the solvent SH by dehydrogenation. This is illustrated below for the case of benzoyl peroxide:

$$(PhCO)_2O_2 \rightarrow 2PhCO_2\cdot \qquad \text{(thermolysis)}$$

$$PhCO_2\cdot + SH \rightarrow PhCO_2H + S\cdot \qquad \text{(abstraction)}$$

$$S\cdot + (PhCO)_2O_2 \rightarrow PhCOOS + PhCO_2\cdot \qquad \text{(induced decomposition)}$$

In the acyl peroxides $(RCO)_2O_2$ in which R is a secondary or tertiary alkyl or a benzyl group, breakdown may also occur in part by a polar mechanism to give the esters RCOOR.

### (A) *Alkyl peroxides*

Among the dialkyl peroxides the most intensively studied is the di-*t*-butyl derivative. This is due mainly to (a) its extensive use as an initiator in homolytic reactions, which follows from its safety in handling compared with its extremely hazardous homologues such as methyl and ethyl peroxides, and (b) its freedom from induced decomposition (however, see below). The peroxide has reportedly been distilled under atmospheric pressure (b.p. 104 °C) with but slight decomposition, and its half-life found to be 11 years at 60 °C, 11 hours at 125 °C, and 35 seconds at 180 °C, so that reactions using this peroxide as the source of radicals can be carried out at convenient temperatures in the neighbourhood of 120 °C. The *t*-butoxy radicals produced on thermolysis may break down further into free methyl and acetone, a disproportionation process which becomes more important with rising temperature:

$$Bu^tO\text{—}OBu^t \rightarrow 2Bu^tO\cdot$$

$$Bu^tO\cdot \rightarrow MeCOMe + Me\cdot$$

The thermal decomposition of *t*-butyl peroxide both in the gas phase and in solution has been well investigated. In the gas phase this proceeds by a first order mechanism, is independent of pressure and over the range 116 °C–350 °C fits a rate constant[1] of $7 \times 10^{15} \exp(-38\,000/RT)$. In cumene, *t*-butylbenzene, or tri-*n*-butylamine as solvent, decomposition of the peroxide occurs at essentially the same rate, indicating that the rate determining step is the simple unimolecular process of bond scission, free from complication by any form of induced chain reaction.[2] This immunity from induced decomposition by solvent radicals, which has made this peroxide in many applications a superior radical source to other peroxides such as benzoyl peroxide, has been accepted for a long time. The only cases of induced decomposition encountered were with radicals derived from primary and secondary alcohols and amines, with which transfer of hydrogen takes place (Huyser)[3a] as illustrated below:

$$R_2\dot{C}-O-H \quad + \quad \begin{matrix} Bu^tO \\ | \\ OBu^t \end{matrix} \longrightarrow R_2C{=}O + Bu^tO\cdot + Bu^tOH$$

or when the peroxide is heated by itself, when abstraction of hydrogen from a methyl group in the peroxide occurs with the ultimate formation of iso-butylene oxide,[3b] as shown:

$$Bu^t_2O_2 \xrightarrow{Bu^tO\cdot} Me_2\overset{\overset{\displaystyle CH_2\cdot}{|}}{C}-O-O-Bu^t \longrightarrow Me_2\overset{\overset{\displaystyle CH_2}{|}}{C}\text{——}O + \cdot OBu^t$$

However, a case of induced decomposition of the $S_H2$ type, as occurs with benzoyl peroxide mentioned earlier, has recently been encountered[4] in the decomposition of di-*t*-butyl peroxide in the benzyl ethers $ArCH_2OMe$, with formation of the mixed acetal $ArCH(OMe)OBu^t$, probably by the following reaction path:

$$Ar\dot{C}HOMe + \begin{matrix} Bu^tO \\ | \\ OBu^t \end{matrix} \longrightarrow ArCH\begin{matrix} OBu^t \\ OMe \end{matrix} + Bu^tO\cdot$$

The extent to which this form of scission of the peroxide bond takes place appears to depend on the substituent in the aryl group Ar. Goh and Ong[5] have studied the decomposition of the peroxide in a series of benzyl ethers under standard conditions (heating in a sealed tube for 72 hours at 110 °C) and have found varying degrees of decomposition of the peroxide, indicating occurrence of reactions by pathways other than the unimolecular homolysis of the peroxidic linkage (see Table 6.1). When extended to a number of other solvents (Table 6.2), the extent of decomposition of the peroxide is again found to vary. Thus, for those solvents such as tri-*n*-butylamine, anisole, *t*-butylbenzene, in which it has been established that no induced decomposition occurs, the degrees of decomposition after the standard treatment

**Table 6.1** Decomposition of *t*-Butyl Peroxide in Benzyl Ethers at 110 °C for 72 hours[5]

| Solvents | % Decomposition of Peroxide | ArCH(OMe)OBu$^t$ (% yield) |
|---|---|---|
| Benzyl phenyl ether | 54 | 0 |
| Benzyl methyl ether | 56 | 3 |
| 4-*t*-Butylbenzyl methyl ether | 60 | 9 |
| 4-Nitrobenzyl methyl ether | 61 | 85 |
| Benzyl ethyl ether | 62 | 0 |
| 4-Chlorobenzyl methyl ether | 85 | 36 |
| 3,4-Dichlorobenzyl methyl ether | 99 | 62 |

**Table 6.2** Percentage Decomposition of *t*-Butyl Peroxide in Various Solvents at 110 °C for 72 hours[5]

| Solvent | % Decomposition |
|---|---|
| Decalin | 43 |
| *NN*-Dimethylbenzylamine | 47 |
| Anisole | 48 |
| Tri-*n*-butylamine | 48 |
| *t*-Butylbenzene | 49 |
| Ethylbenzene | 54 |
| 4-Chlorotoluene | 59 |
| Benzyl chloride | 60 |
| Benzyl acetate | 60 |
| Benzylamine | 62 |
| Benzyl cyanide | 65 |
| Benzaldehyde | 68 |
| Chlorobenzene | 69 |
| Carbon tetrachloride | 71 |
| Carbon tetrachloride | 58* |
| 4-Chlorobenzyl cyanide | 72 |
| Tetrachloroethylene | 83 |
| Benzyl alcohol | 85 |
| α-Methylbenzyl alcohol | 88 |

* In the presence of 0.1 M iodine.

are all around 48%, whereas in some other solvents, notably the benzyl cyanides and alcohols, this figure is much exceeded. In the case of the benzyl alcohols this excess over 48% may in whole or in part be accounted for by induced decomposition through hydrogen transfer, as mentioned above; in the other cases it may well be that induced decomposition of the type under consideration has occurred.

### (B) *Acyl peroxides*

The acyl peroxides whose thermal homolysis has been the most intensively studied are the acetyl and benzoyl derivatives.

Thermolysis of acetyl peroxide when conducted in solution (inert solvents) leads to more complex products than when carried out in the gas phase, which gives mainly free methyl radicals (ultimately appearing as ethane) and carbon dioxide. This is due to the occurrence, in solution, of 'cage reactions' (see also Section 5.5) as indicated in the following chart, in which 'caged' radicals are shown within square brackets:

$$\text{MeCO—O—O—COMe} \begin{array}{l} \nearrow \; 2\text{Me}\cdot + 2\text{CO}_2 \\ \searrow \; [\text{MeCOO}\cdot \quad \cdot\text{OCOMe}] \end{array}$$

$$[2\text{MeCOO}\cdot] \rightarrow 2\text{MeCOO}\cdot$$

$$\text{MeCOO}\cdot \rightarrow \text{Me}\cdot + \text{CO}_2$$

$$2\text{Me}\cdot \rightarrow \text{C}_2\text{H}_6$$

$$[2\text{MeCOO}\cdot] \rightarrow \text{MeCOOMe} + \text{CO}_2$$

$$[2\text{MeCOO}\cdot] \rightarrow \text{C}_2\text{H}_6 + 2\text{CO}_2$$

The occurrence of the recombination reaction of acetoxy radicals has been confirmed by tracer experiments (Pryor and Smith)[6] using acetyl peroxide in which the carbonyl oxygen is labelled with $^{18}$O. These show that at 80 °C in isooctane 38% of the geminate radicals recombine to form the peroxide. This is in fact the most important cage reaction the acetoxy radicals undergo, since the yields of ester and ethane are only about 15% and 5%. When the labelled peroxide was decomposed in cyclohexene, the label was completely randomized in the cyclohexyl acetate produced, showing that this ester is not a cage product. An acyloxy radical-solvent complex is considered to be the only form in which the acetoxy radical escapes the solvent cage.

Benzoyl peroxide, probably the most widely used reagent in radical reactions on account of its relative convenience and safety in handling and its moderate decomposition temperature, has a half-life at 100 °C of approximately 30 minutes. Thermolysis gives two benzoyloxy radicals, which, unless trapped by some reactive substance present in the reaction system, break down further to the free phenyl radical and carbon dioxide.

$$\mathrm{PhCOO\cdot \rightarrow Ph\cdot + CO_2}$$

That the benzoyloxy radical is first formed is shown by the fact that, when thermolysis is carried out in moist carbon tetrachloride containing iodine, benzoic acid becomes the final product (Hammond).[7] Iodine acts here as a 'scavenger' and presumably reacts with the benzoyloxy radical as follows:

$$\mathrm{PhCOO\cdot \xrightarrow[I_2]{} PhCOOI \xrightarrow[H_2O]{} PhCOOH}$$

More direct evidence has subsequently been found in the decomposition of the peroxide in dibenzyl ether, an effective hydrogen donor, from which over 80 % of benzoic acid was isolated.[8]

$$\mathrm{PhCOO\cdot + PhCH_2{-}O{-}CH_2Ph \rightarrow PhCOOH + Ph\dot{C}H{-}O{-}CH_2Ph}$$

The decomposition mechanism of benzoyl peroxide has received exhaustive study and has been found to be very complex (see also Section 13.3). This is due largely to the fact that in addition to homolysis of the peroxidic linkage, an induced chain decomposition of the peroxide also takes place. This side reaction, which may assume serious proportions unless reaction conditions are strictly controlled, was first deduced from the observation that the apparently unimolecular rate for decomposition of the peroxide increases with peroxide concentration, and is also increased by the addition of other radical sources and decreased by the introduction of oxygen or similar inhibitors such as quinones and polynitroaromatics. By using isotopic oxygen, it has been possible to demonstrate[9a] that attack by a solvent radical S· occurs on one of the peroxidic oxygen atoms, the key step, it is believed, being transfer[9b] of an electron from S·, as follows:

$$\mathrm{S\cdot + \begin{matrix} PhCO & & COPh \\ & \diagdown\ \ \diagup & \\ & O{-}O & \end{matrix} \rightarrow PhCOOS + \cdot OCOPh}$$

Understandably, the extent to which induced decomposition occurs increases with the concentration of benzoyl peroxide. Thus benzoyl peroxide follows a first order law in its decomposition only in very dilute solutions.

When the peroxide is heated in the absence of any solvent the products are carbon dioxide, biphenyl and small amounts of phenyl benzoate and benzene. At high dilution in benzene, the main products remain carbon dioxide and biphenyl, but other products resulting from the attack of phenyl radicals on the solvent are also found. For example, decomposition[10] of an 0.025 M solution in benzene at 80 °C proceeds as follows:

$$(PhCO)_2O_2 + PhH \rightarrow CO_2 + PhCO_2H + Ph{-}Ph +$$

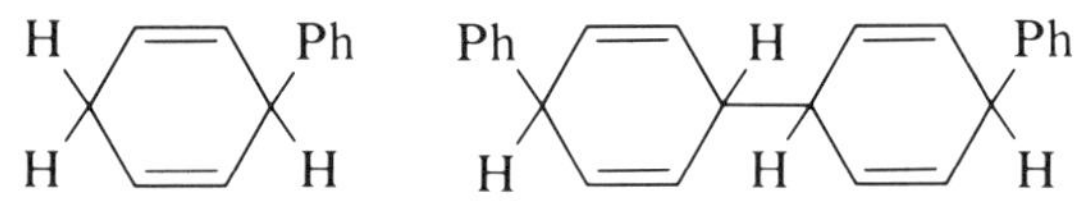

H H Ph H      Ph H H H Ph H

In substituted benzoyl peroxides electron attracting substituents (*m*- and *p*-Cl, *m*- and *p*-CN, etc.) slow down decomposition, whereas electron donating groups (*m*- and *p*-Me or *p*-*t*-Bu) accelerate it. This has been explained as being due to the effect of the substituents on the dipole interaction across the peroxidic bond, electron attracting substituents diminishing the magnitude of this interaction and thus lowering the rate of decomposition, and vice versa, as shown.

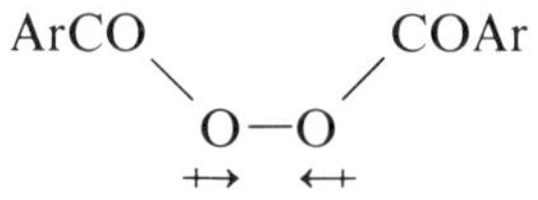

ArCO COAr
O—O
$\leftrightarrow$ $\leftrightarrow$

(C) *Other peroxy compounds*

Thermolysis of *t*-butyl hydroperoxide, the most important source of radicals among the hydroperoxides, gives *t*-butoxy and hydroxy radicals. Cumyl hydroperoxide, another useful member of the group, decomposes (at around 100 °C) as follows:

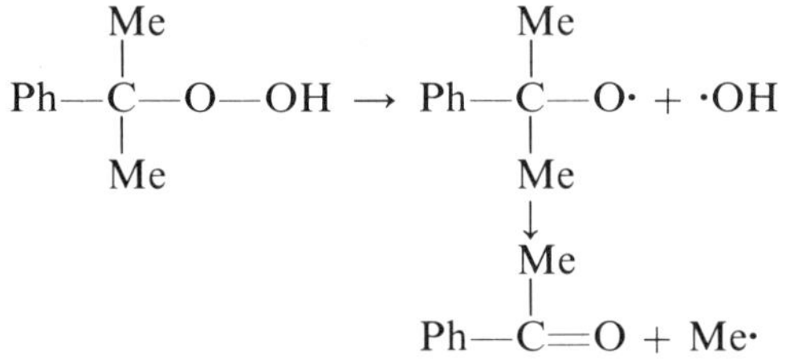

$$Ph{-}C(Me)_2{-}O{-}OH \rightarrow Ph{-}C(Me)_2{-}O\cdot + \cdot OH$$

$$\downarrow$$

$$Ph{-}C(Me){=}O + Me\cdot$$

Among peresters perhaps the best known is *t*-butylperbenzoate, which on heating generates a benzoyloxy and a *t*-butoxy radical:

$$PhCOO{-}OBu^t \rightarrow PhCOO\cdot + \cdot OBu^t$$

Another perester, di-*t*-butyl peroxyoxalate ($Bu^tO—O_2C—)_2$, is a specially useful source of *t*-butoxy radicals in that it suffers thermal decomposition at relatively low temperatures (30–45 °C). The carbon dioxide which it liberates on thermolysis incidentally increases the efficiency of radical production as compared to *t*-butyl peroxide, by aiding diffusion of the radicals from the solvent cage and thus cutting down cage reactions. An interesting example of the use of this initiator is the generation of the *o*-benzoylbenzyl radical (1) from *o*-benzoyltoluene at 40 °C, at which temperature the radical dimerizes to 2,2′-dibenzoyldibenzyl, whereas using *t*-butyl peroxide would require a higher temperature (120–125 °C), at which the radical produced undergoes intramolecular alkylation to give anthrone which in turn suffers dehydrogenation to anthraquinone and dianthrone (Huang and Lee),[11] as shown.

COPh, $CH_3$ —t-Bu$^t$O·→ CO, $CH_2$· (1) —40 °C→ dimer; —120 °C→ CO, $CH_2$ → anthraquinone + [anthrone]$_2$

## 6.2 Azo compounds

Thermal decomposition of the azo compounds R—N=N—R′, some of which are much employed sources of radicals, gives the radicals R· and R′· and invariably also nitrogen, the formation of which (heat of formation 225 kcal mole$^{-1}$) supplies a strong driving force for the dissociation process:

$$R—N{=}N—R' \rightarrow R\cdot + N_2 + R'\cdot$$

From data on some of the better known compounds, listed in Table 6.3, it is seen that the ease of decomposition depends on the stability of the radical R· and/or R′· being generated. Thus the decomposition of azomethane requires an activation energy of 53 kcal mole$^{-1}$ and occurs at appreciable

**Table 6.3** Decomposition of Azo Compounds*

| Compound | Solvent | Temp. °C | $k \times 10^4$ ($s^{-1}$) | $E_\alpha$ (kcal $mole^{-1}$) |
|---|---|---|---|---|
| Me—N=N—Me | gas | 300 | 5.6 | 52.7 |
| $Me_2CH$—N=N—$CHMe_2$ | gas | 250 | 4.8 | 40.9 |
| Ph—N=N—$CPh_3$ | benzene | 25 | 0.042 | 26.8 |
| Ph—N=N—$CHPh_2$ | decalin | 54 | 1.01 | 34.0 |
| PhMeCH—N=N—CHMePh | toluene | 110 | 1.69 | 32.6 |
| $Ph_2CH$—N=N—$CHPh_2$ | toluene | 64 | 3.40 | 26.6 |

* Data from C. Walling, *Free Radicals in Solution*, John Wiley, New York, 1957.

rates only at temperatures near 300 °C. On the other hand phenylazotriphenylmethane undergoes homolysis between 45 °C to 80 °C, the activation energy here being only 27 kcal $mole^{-1}$, presumably because the triphenylmethyl radical to be generated is resonance stabilized. It is considered that in the transition state the two C—N bonds are broken simultaneously.[13]

The effects of nuclear substituents R on the rate of thermal decomposition of the azocumenes (2) have been probed[14] and the results reveal that the rates are all first order, and that the influences of R are small but significant and best explained on the basis of resonance (including hyperconjugative) contributions outweighing the inductive effect. Thus the effect of R on the ease of decomposition decreases in the order Br > Me > Et > $Pr^i$, $Bu^t$, and H.

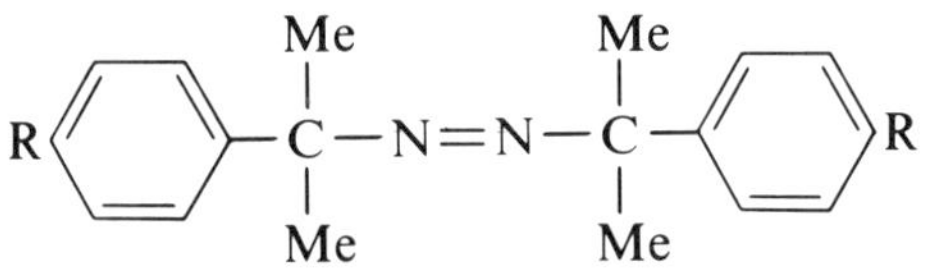

$$R\text{-}C_6H_4\text{-}C(Me)_2\text{-}N{=}N\text{-}C(Me)_2\text{-}C_6H_4\text{-}R$$

(2)

There has been considerable interest in cyclic azo compounds. Overberger and Lapkin[15] studied the decomposition of the compound (3) in inert solvents and obtained the product (4), presumably via the generation of a pair of diradicals which then dimerized.

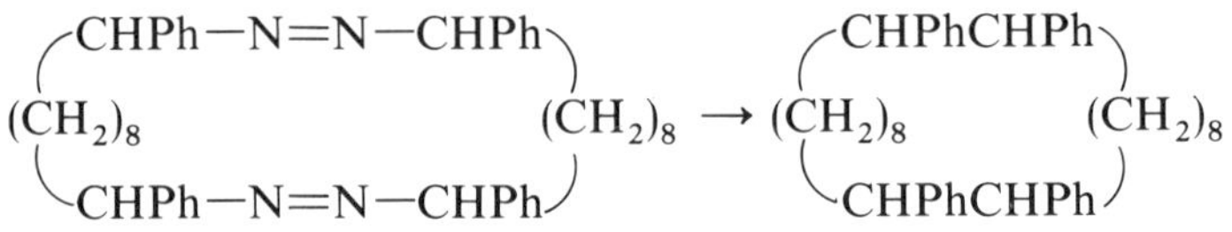

$$\begin{matrix} \text{CHPh—N=N—CHPh} \\ (CH_2)_8 \qquad\qquad (CH_2)_8 \\ \text{CHPh—N=N—CHPh} \end{matrix} \rightarrow \begin{matrix} \text{CHPhCHPh} \\ (CH_2)_8 \qquad (CH_2)_8 \\ \text{CHPhCHPh} \end{matrix}$$

(3) (4)

The thermolysis of 2,3-diazonorbornene derivatives[16] has been found to produce bicyclopentanes with predominant inversion at the bridgehead positions. It has been suggested that the inverted diradicals (see scheme below) are formed following nitrogen elimination, and that this inversion is due to the recoil energy released by C—N bond breaking. Cyclization then competes with further inversion.

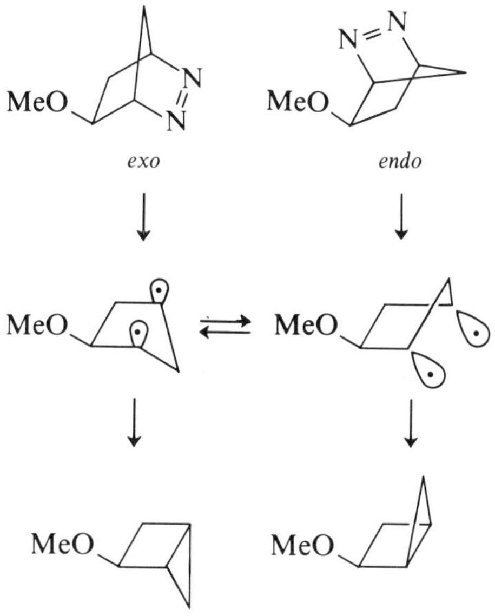

*exo* *endo*

*trans* *cis*

37% from *exo* 63% from *exo*
6.4% from *endo* 93.6% from *endo*

The thermolysis of azo *bis* isobutyronitrile (AIBN), and of benzenediazoacetate which was one of the earliest among azo compounds to be studied in connection with radical reactions, are discussed elsewhere (see Sections 5.5 and 13.1, respectively).

The particularly facile thermolysis of *t*-butyl hyponitrite,[17] which occurs at about 60 °C, to give *t*-butoxy radicals with evolution of nitrogen:

$$\mathrm{Bu^tO{-}N{=}N{-}OBu^t} \rightarrow \mathrm{2Bu^tO\cdot} + \mathrm{N_2}$$

renders the reagent a more efficient radical source than the peroxide and, like the peroxyoxalate, enables study to be made of certain reactions at lower temperatures than is possible when the peroxide is employed. Thus the hyponitrite has been used in an experiment to show that the α-methoxybenzyl radical $\mathrm{Ph\dot{C}HOMe}$ is capable of abstracting a chlorine atom from carbon tetrachloride to form the α-chloro-ether PhCHCl(OMe) in the following chain reaction in which carbon tetrachloride has been assigned[18] the role of a chlorinating agent:

$$PhCH_2OMe + Bu^tO\cdot \rightarrow Ph\dot{C}HOMe + Bu^tOH$$

$$Ph\dot{C}HOMe + CCl_4 \rightarrow PhCHCl(OMe) + \cdot CCl_3$$

$$\cdot CCl_3 + PhCH_2OMe \rightarrow Ph\dot{C}HOMe + CHCl_3 \ldots \text{etc.}$$

When the experiment was carried out at the higher decomposition temperature of the peroxide (120 °C) the α-chloro-ether could not be isolated, as it is unstable at this temperature and breaks down to benzaldehyde and methyl chloride.

In another instance, the hyponitrite has been utilized to generate, in the presence of *t*-butyl peroxide, α-methoxybenzyl radicals at 60 °C (at which temperature the peroxide is stable), which then induces the decomposition of the peroxide to give the mixed acetal PhCH(OMe)OBu$^t$, as discussed earlier.[4]

## 6.3 Organometallics

The homolysis of the carbon–metal bond occurs readily in a number of organometallic compounds. Thermolysis of tetramethyl lead, in fact, furnished one of the earliest demonstrations of the transient existence of alkyl radicals (Section 2.4). From the mean bond dissociation energies of a number of metal alkyls, given in Table 6.4, the relatively low values found for organomercurials imply that the carbon–mercury bond is relatively susceptible to thermal cleavage and hence that this group of organometallic

**Table 6.4** Mean Bond Dissociation Energies of Certain Organometallic Compounds

| Compound | kcal mole$^{-1}$ | Compound | kcal mole$^{-1}$ |
|---|---|---|---|
| Hg–Pr$^i$ | 21 | Zn–Et | 35 |
| Hg–Et | 24 | Zn–Me | 42 |
| Hg–Me | 29 | Sn–Me | 52 |
| Hg–Ph | 32 | Sn–Et | 46 |
| Pb–Et | 31 | Li–Et | 50 |
| Pb–Me | 37 | Li–Bu | 59 |

From H. A. Skinner, The Strengths of Metal-to-Carbon Bonds, *Adv. Organometal. Chem.*, 1964, p. 102. (The mean bond dissociation energy is derived from the heat of formation of a compound $MR_n$ divided by n and is different from bond dissociation energy.)

compounds are ready sources of free radicals, as indeed they are. They suffer, however, from the disadvantage of being highly toxic, and this greatly limits their use in industry or in the laboratory.

If metal alkyls yield radicals on heating, unstable organometallic complexes would yield radicals also. Thus alkylchromium complexes have been shown to decompose homolytically while several other transition metal complexes are known to initiate polymerization, presumably in the same way or else by redox pathways. Grignard reagents, which normally react by an ionic mechanism, follow a totally different reaction path under the influence of catalytic quantities of certain metal halides, such as cobaltous chloride. The reaction has been extensively studied by Kharasch and his co-workers,[19] and while it is clear that radicals are involved, the exact nature of the reaction has not yet been ascertained. It may be classified as a redox process although the formation of organocobalt intermediates followed by their homolytic decomposition cannot be ruled out. Lead tetra-acetate, a strong oxidizing agent, under certain conditions decomposes to yield acetoxy radicals although its reaction via ionic pathways is more common. The thermal decomposition of organolithiums is apparently homolytic in nature, and in several instances radical formation has been detected from the observed polarization of nuclear magnetic resonance spectra (see Section 2.3).

Except for a limited number of special applications, the organometallic compounds discussed above have found much fewer uses than the peroxides or the azo compounds. During recent years, however, a group of organotin compounds has come into prominence by virtue of their versatility and convenience in usage. These are the organotin hydrides, reactions of which include the reduction of alkyl or aryl halides and radical chain addition to olefins.[20] These reactions can be initiated thermally, photolytically or by added initiators and are illustrated below (trialkyltin hydride being represented by $R'_3SnH$ and the initiating radical by In·).

*Reduction of halides RX by organotin hydride*

$$\text{Initiator} \longrightarrow 2\text{In}\cdot$$

$$\text{In}\cdot + R'_3SnH \longrightarrow R'_3Sn + \text{InH}$$

$$R'_3Sn + RX \longrightarrow R\cdot + R'_3SnX$$

$$R\cdot + R'_3SnH \longrightarrow RH + R'_3Sn$$

$$2R'_3Sn \longrightarrow R'_3Sn\text{—}SnR'_3$$

$$R\cdot + R'_3Sn \longrightarrow RR'_3Sn$$

$$R\cdot \longrightarrow \text{abstraction, disproportionation or coupling products}$$

*Addition of organotin hydride to olefins*

$$\text{Initiator} \rightarrow 2\text{In}\cdot$$

$$\text{In}\cdot + R'_3\text{SnH} \rightarrow R'_3\text{Sn} + \text{InH}$$

$$R'_3\text{Sn} + \text{>C=C<} \rightarrow R'_3\text{Sn—}\overset{|}{\underset{|}{\text{C}}}\text{—C}\cdot\text{<}$$

$$R'_3\text{Sn—}\overset{|}{\underset{|}{\text{C}}}\text{—C}\cdot\text{<} + R'_3\text{SnH} \rightarrow R'_3\text{Sn—}\overset{|}{\underset{|}{\text{C}}}\text{—}\overset{|}{\underset{|}{\text{C}}}\text{—H} + R'_3\text{Sn}$$

The addition reaction as can be seen, has a mechanism quite similar to that of hydrogen bromide. Its radical nature is evidenced by (a) catalysis by azo *bis* isobutyronitrile, oxygen or ultraviolet irradiation, and (b) inhibition by radical scavengers.

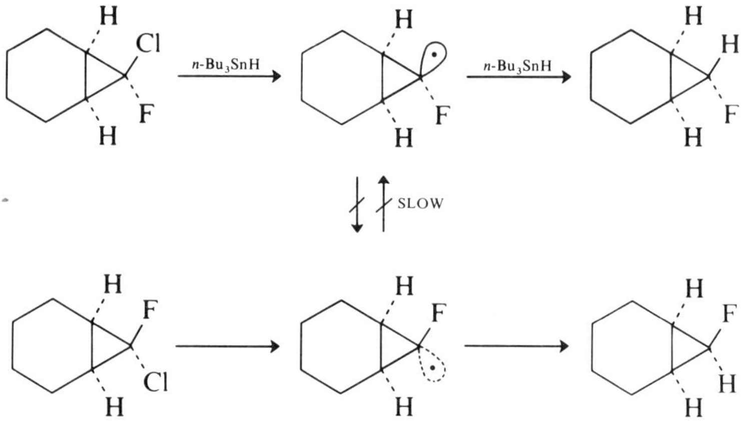

An interesting illustration of the use of organotin hydrides is found in the reduction of *gem*-chlorofluorocyclopropanes, which proceeds with complete stereospecificity.[21] Two features are noteworthy in this reaction. First, the observed 100% stereospecificity of the reduction requires a very rapid rate of hydrogen transfer to the intermediate cyclopropyl radical and this indicates that trialkyltin hydrides are extremely reactive towards radicals. Second, while most radicals are flat, the substituent effect of fluorine apparently helps to stabilize the pyramidal structure of the radical, rendering inversion slower than hydrogen transfer.

## References


1 F. LOSSING and A. W. TICKNER, *J. chem. Phys.*, 1952, **20**, 907.
2 J. H. RALEY, F. F. RUST, and W. E. VAUGHAN, *J. Am. chem. Soc.*, 1948, **70**, 1336.
3*a* E. S. HUYSER and C. J. BREDEWEG, *J. Am. chem. Soc.*, 1964, **86**, 2401.
*b* E. R. BELL, F. F. RUST, and W. E. VAUGHAN, *J. Am. chem. Soc.*, 1950, **72**, 337.
4 R. L. HUANG, T. W. LEE, and S. H. ONG, *J. chem. Soc.* (*C*), 1969, 2522.
5 S. H. GOH and S. H. ONG, *J. chem. Soc.* (*B*), 1970, 870.
6 W. A. PRYOR and K. SMITH, *J. Am. chem. Soc.*, 1967, **89**, 1741.
7 G. S. HAMMOND and L. M. SOFFER, *J. Am. chem. Soc.*, 1950, **72**, 4711.
8 R. L. HUANG, H. H. LEE, and S. H. ONG, *J. chem. Soc.*, 1962, 3336.
9*a* D. B. DENNY and G. FEIG, *J. chem. Soc.*, 1959, **81**, 5322.
*b* See M. J. PERKINS, *Annual Reports of the Progress of Chemistry* (*B*), 1969, 168.
10 D. F. DETAR and R. A. J. LONG, *J. Am. chem. Soc.*, 1958, **80**, 4742; D. H. HEY, M. J. PERKINS, and G. H. WILLIAMS, *J. chem. Soc.*, 1964, 3412; G. B. GILL and G. H. WILLIAMS, *J. chem. Soc.*, 1965, 995.
11 R. L. HUANG and H. H. LEE, *J. chem. Soc.* (*C*), 1966, 929.
12 See C. WALLING, *Free Radicals in Solution*, John Wiley, New York, 1957, p. 516.
13 S. SELTZER and S. G. MYLONAKIS, *J. Am. chem. Soc.* 1967, **89**, 6584.
14 J. R. SHELTON, C. K. LIANG, and P. KOVACIC, *J. Am. chem. Soc.*, 1968, **90**, 354.
15 C. G. OVERBERGER and M. LAPKIN, *J. Am. chem. Soc.*, 1955, **77**, 4651.
16 E. ALLRED and R. C. SMITH, *J. Am. chem. Soc.*, 1967, **89**, 7133.
17 H. KIEFER and T. G. TRAYLOR, *Tetrahedron Lett.*, 1966, 6164.
18 R. L. HUANG, T. W. LEE, and S. H. ONG, *J. chem. Soc.* (*C*), 1969, 40.
19 M. S. KHARASCH and E. K. FIELDS, *J. Am. chem. Soc.*, 1941, **63**, 2316; M. S. KHARASCH and R. L. HUANG, *J. org. Chem.*, 1952, **17**, 669; for other references see W. A. WATERS Ed., *Vistas in Free Radical Chemistry*, Pergamon, London, 1959, pp. 14–15.
20 H. G. KUIVILA, *Acc. Chem. Res.*, 1968, **1**, 299.
21 T. ANDO, F. NAMIGATA, H. YAMANAKA, and W. FUNASAKA, *J. Am. chem. Soc.*, 1967, **89**, 5719.

# 7

# Photochemical Processes

Although the profound effect of light on organic compounds has been known for well over a hundred years, the majority of the more spectacular advances in this field of study were not made until the past decade. This field, however, properly belongs to the realm of photochemistry and the discussion given here will be confined to the generation of radicals and to certain radical reactions of special interest, rather than the detailed mechanisms of photochemical processes. The photolysis of chemical compounds not only provides a convenient method for the generation of radicals through bond homolysis but also opens up a new chemistry, that of molecules in the excited state.

Visible light of 500 nm or 5000 Å wavelength is equivalent to 57.2 kcal mole$^{-1}$ calculated according to the equation $E = h\nu$, while ultraviolet (UV) radiation of 300 nm or 3000 Å corresponds to 95.3 kcal mole$^{-1}$. An examination of the Table of bond dissociation energies given in Section 3.2 will therefore show that UV light of this wavelength and less (energies of $>$95 kcal mole$^{-1}$) is sufficient to cause, among other things, the homolysis of almost all types of single bond in organic compounds. However, for a photochemical reaction to occur, the radiation must first be absorbed by the molecule, and whether this situation obtains can be ascertained from the absorption bands exhibited which correspond to the wavelengths of the incident radiation that are absorbed.

## 7.1 Basic theory of photochemical processes[1]

### (A) *Excitation and de-excitation*

Organic molecules which absorb UV and visible radiation are usually those containing $\pi$-bonds or non-bonding electron pairs. One of the best studied is the carbonyl bond which has $\pi$-bonding ($\pi$) and antibonding ($\pi^*$) orbitals as well as non-bonding ($n$) orbitals of lone electron pairs on the oxygen atom. Consequently the carbonyl function gives rise to two absorption maxima, a strong band at *ca* 190 nm with an extinction coefficient of *ca* $10^3$ due to a ($\pi$–$\pi^*$) transition and a weak band at *ca* 280 nm with extinction of *ca* 10 due to a ($n$–$\pi^*$) transition. A pictorial representation is given in Fig. 7.1. The transitions, whether ($\pi$–$\pi^*$) or ($n$–$\pi^*$), are governed by quantum restrictions, one of which states that singlet-triplet transitions are forbidden. This means that the process of excitation from the ground state to an excited state occurs with total electron spin retention (or conservation). Since the ground state for most organic molecules (except for radical or diradical species) is necessarily singlet, i.e., all spins are paired,

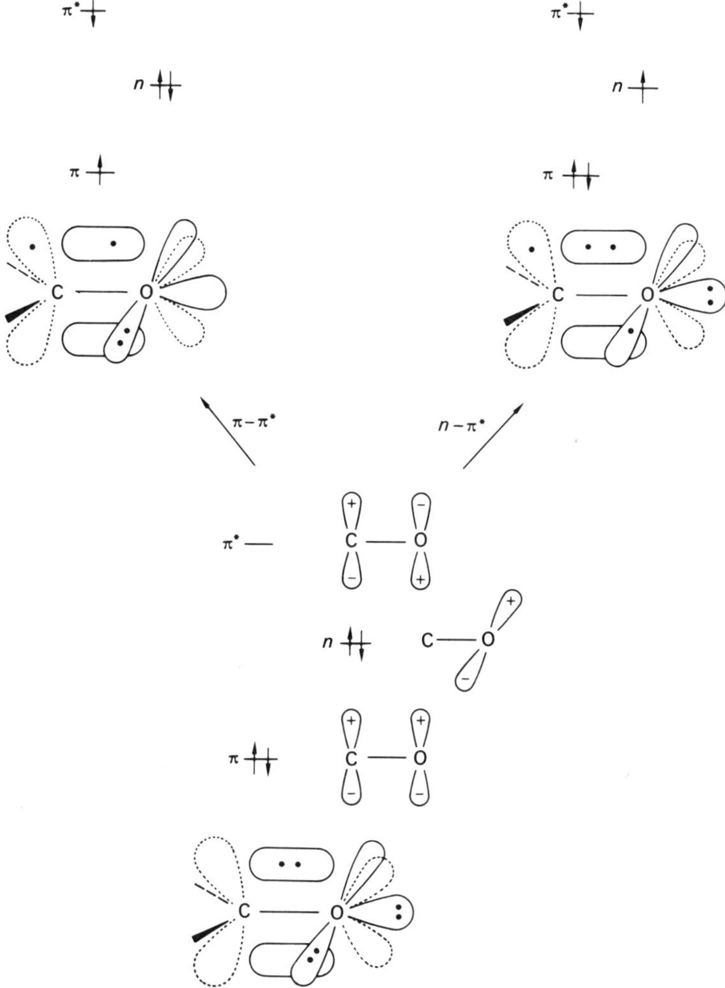


**Fig. 7.1** Orbitals and electronic transitions of the carbonyl group.

only excitation from singlet ground states to excited singlet states can occur and direct excitation to triplet states is forbidden.

Furthermore, the electronic transition from the ground state to the excited state is governed by the Franck–Condon principle which states that the electronic changes during light absorption are so rapid that there is no change in the positions of the atomic nuclei. Thus for any molecule A–B with potential energy curves as shown in Fig. 7.2, the transition is essentially

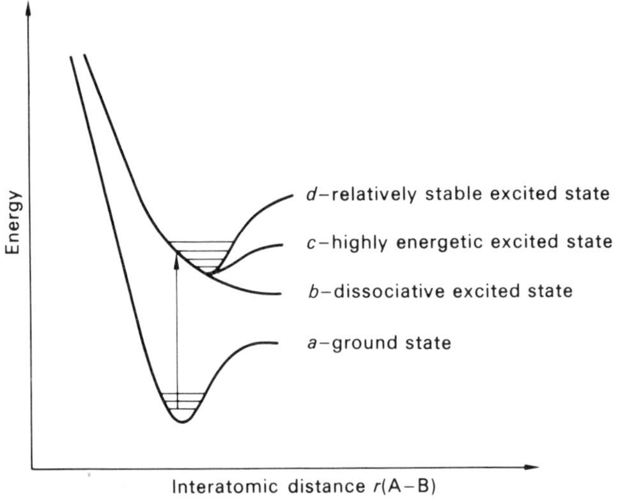


**Fig. 7.2** Franck–Condon excitation to three types of excited states.

vertical, i.e., the interatomic distance $r$(A–B) remains constant. If the transition is from the ground state (represented by curve *a*) to a dissociative excited state (curve *b*), the molecule after absorption of light dissociates or decomposes to radical fragments A· and B·. This behaviour is shown by a large number of bond types and the radicals so generated can be used directly for initiation of chain reactions. A transition to potential curve *c* leads to an excited state which is energetic in character and can similarly result in dissociation whereas a transition to potential curve *d* gives a relatively stable excited state with a longer lifetime.

A molecule in an excited state, produced by absorption of a quantum of light, may lose the absorbed energy and return to the ground state by a number of physical processes, *viz*., (a) internal conversion, (b) intersystem crossing, (c) fluorescence, (d) phosphorescence or (e) energy transfer. Alternatively, and this is of greater interest from the chemical standpoint, the excited molecule may undergo one or more of a variety of reactions. For a pictorial representation of physical processes of excitation and de-excitation, one is referred to the Jablonski diagram given in Fig. 7.3. An allowed vertical transition occurs from the singlet ground state $S_0$ to either excited states $S_1$ or $S_2$ or to any of their vibration-rotational levels, and through a series of radiationless transitions called internal conversions the excited molecule drops to the $S_1$ state with liberation of heat energy. The molecule may decay to the ground state $S_0$ by undergoing either fluorescence with emission of

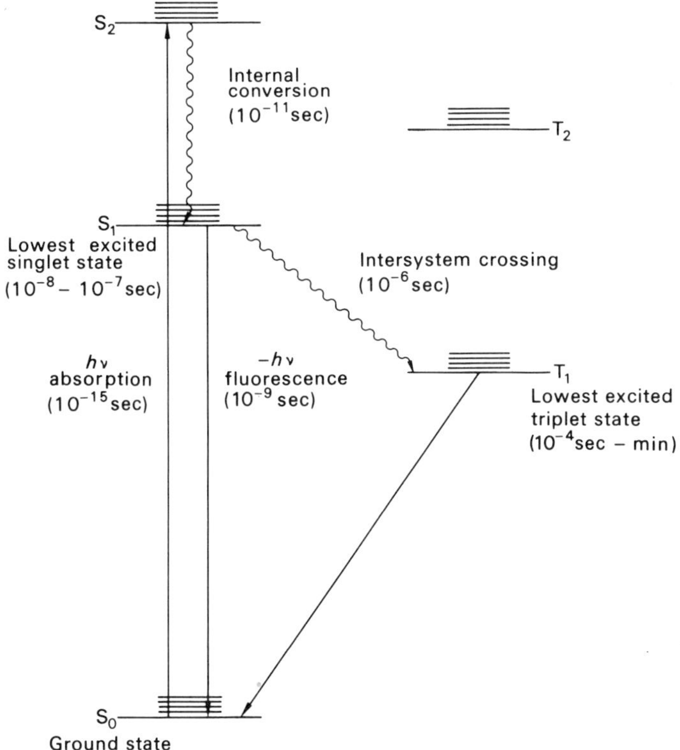


**Fig. 7.3** Jablonski Diagram (lifetimes and rates given are only approximate).

radiation or radiationless intersystem crossing to the triplet state $T_1$, after which phosphorescence to $S_0$ can occur with emission of light. Pure singlet-triplet transitions are quantum mechanically forbidden but since real states are not pure singlets or triplets but can have mixed with them states of different multiplicities, such transitions while not totally forbidden are of low probability. The triplet state thus has a decidedly longer lifetime ($10^{-4}$ seconds to several minutes) than the singlet and due to this relatively longer lifetime it is in the triplet state that most photochemical reactions occur.

(B) *Energy transfer*[2]

Of considerable theoretical interest as well as practical importance is the phenomenon of energy transfer in which an excited molecule, referred to as a donor or photosensitizer, transfers its energy to an acceptor substrate. If the donor is in the triplet state, then triplet-triplet energy transfer will give acceptor triplets. Singlet-singlet energy transfer can also occur but is less common. Acceptor molecules may also be referred to as quenchers, there being triplet quenchers as well as singlet quenchers. (Quenching refers to the reduction of radiative emission processes, i.e., fluorescence and phosphorescence, which occurs when there is energy transfer).

Triplet-triplet energy transfer is particularly useful for the generation of triplet molecules which are difficult to form either because of an inefficient intersystem crossing or if the substrate fails to absorb the irradiating light. One widely used sensitizer is benzophenone, which is both effective and convenient for the following reasons: (a) the intersystem crossing is efficient, (b) light is absorbed at a reasonably high wavelength (360 nm) which is not 'cut' (i.e., filtered) by pyrex glassware, (c) the triplet state is fairly long-lived (*ca* $10^{-5}$ s) and (d) the lowest triplet state is of fairly high energy (*ca* 69 kcal).

Triplet-triplet energy transfer is illustrated in Fig. 7.4. Irradiation allows the excitation of the photosensitizer $S_0$ to its excited singlet state $S_1$ but not of the acceptor $A_0$ to $A_1$ because of the filter cut-off or the higher energy required. Rapid intersystem crossing gives the triplet state $S_3$ and triplet-triplet energy transfer gives the acceptor triplet $A_3$ which can undergo chemical reaction or decay to $A_0$ via phosphorescence.

Simple olefins and dienes are difficult to excite to the triplet state because of a high first excited singlet level as well as an inefficient singlet-triplet

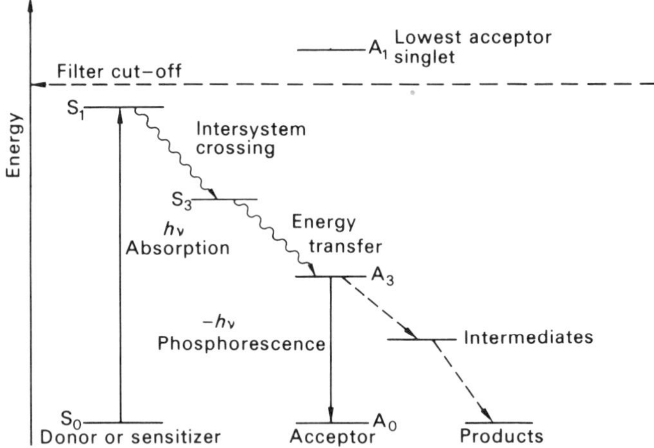


**Fig. 7.4** Triplet-triplet energy transfer.

intersystem crossing, but because of the presence of a low energy triplet state they are excellent triplet acceptors (or triplet quenchers). Oxygen, a ground state triplet, is also a good triplet state quencher, its action being analogous to radical scavenging.

(C) *Quantum yield*

In a photochemical reaction one would be interested in the overall yield of products in relation to the amount of light absorbed. This is expressed as the quantum yield $\Phi$, defined as

$$\Phi = \frac{\text{No. of moles of product (or reactant consumed)}}{\text{No. of einsteins (moles) of light absorbed}}$$

Quantum yields vary widely, and may be as low as $10^{-2}$ to as high as $10^6$. If $\Phi = 0$ all the absorbed energy is lost in physical processes not involving chemical reaction. When $\Phi = 1$, the absorbed quanta are transformed stoichometrically into products. When $\Phi > 1$ it means photo-induced chain reactions have occurred; in some chain reactions such as radical halogenation and polymerization $\Phi$ can be as high as $10^4$–$10^6$.

## 7.2 Photochemistry of carbonyl compounds

Before considering the photochemical reactions of aldehydes and ketones it is instructive to refer to the electronic structure[3] of the carbonyl group (Fig. 7.1). Two types of excitations are possible, *viz.*, ($\pi$–$\pi^*$) and ($n$–$\pi^*$), but since the greater part of the photochemistry of simple carbonyl compounds is studied with commercial mercury arc lamps in pyrex glass vessels transparent to radiation in the ($n$–$\pi^*$) absorption band at about 3200 Å, most of the reactions may be attributed to the ($n$–$\pi^*$) excited states. The ($n$–$\pi^*$) excitation process involves the promotion of one of the non-bonding electrons to the $\pi^*$ orbital. The antibonding $\pi^*$ orbital is more localized on carbon than on oxygen and is shown diagrammatically in Fig. 7.1. This being so means that the electron density on the carbon atom will be higher in the excited state than in the ground state and for the same reason the oxygen atom will be electron-deficient in the excited state. For example, the formaldehyde molecule has the ground state dipole moment of 2.34D reduced to 1.56D in the excited state. In fact, in hydrogen abstraction or hydrogen transfer reactions ($n$–$\pi^*$) excited states of carbonyl function behave as electron-deficient (electrophilic) species, not unlike alkoxy radicals.

The ($n$–$\pi^*$) excited molecule can react in the singlet state with the electrons having paired (*anti*-parallel) spins or in the triplet state with two unpaired

$$\rangle C{=}\ddot{O}: \xrightarrow{n-\pi^*} \rangle \overset{\delta^-}{\dot{C}}{=}\overset{\delta^+}{\dot{O}}:$$

electron spins. The triplet state is usually the reactive state in photochemical reactions of carbonyl compounds although in a few cases the singlet state may be involved. It is therefore not surprising that most radical-like reactions such as hydrogen abstraction are due to triplet states though this does not imply that singlet states cannot show radical-like activity. The difference in

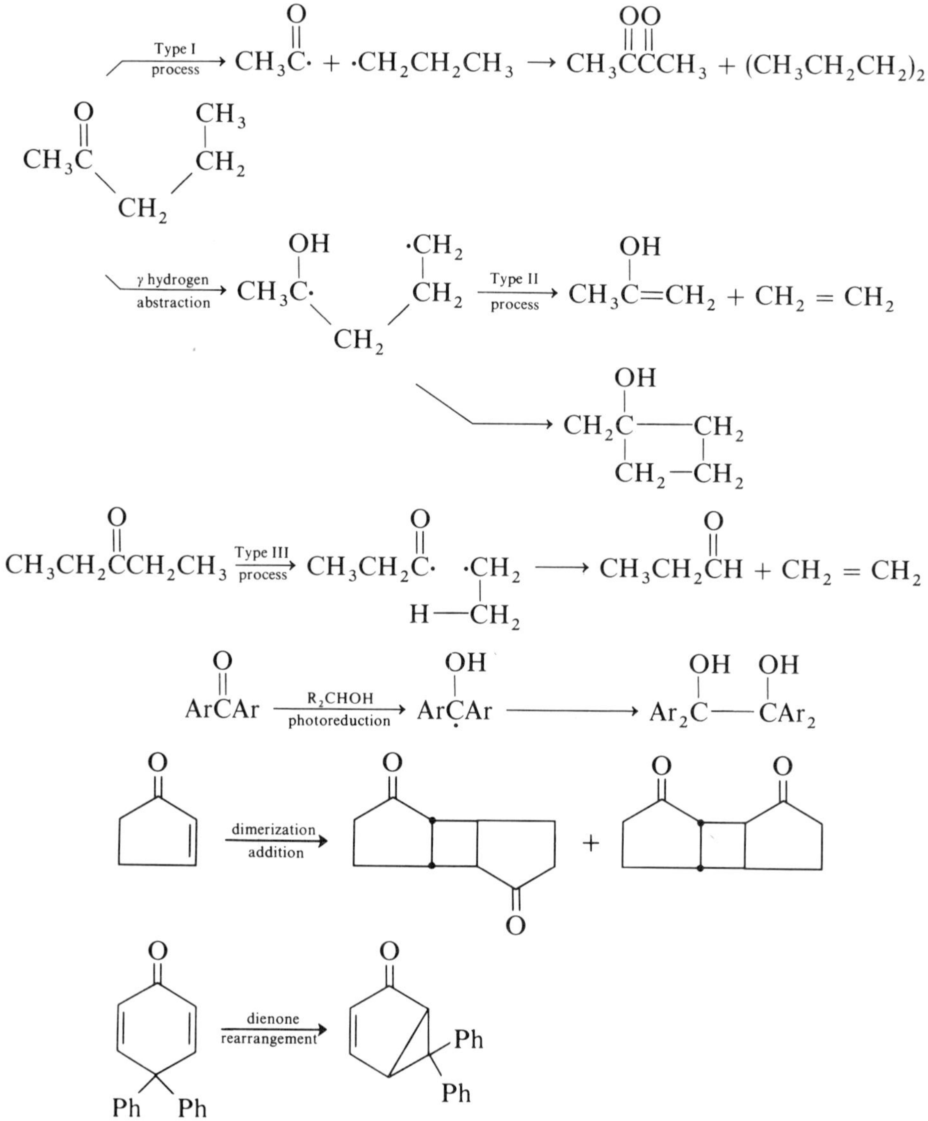

$$CH_3\overset{O}{\overset{\|}{C}}CH_2CH_2CH_3 \xrightarrow[\text{process}]{\text{Type I}} CH_3\overset{O}{\overset{\|}{C}}\cdot + \cdot CH_2CH_2CH_3 \rightarrow CH_3\overset{OO}{\overset{\|\,\|}{CC}}CH_3 + (CH_3CH_2CH_2)_2$$

$$CH_3\overset{O}{\overset{\|}{C}}CH_2CH_2CH_3 \xrightarrow[\text{abstraction}]{\gamma\text{ hydrogen}} CH_3\overset{OH}{\overset{|}{C}}\cdot\ CH_2CH_2\dot{C}H_2 \xrightarrow[\text{process}]{\text{Type II}} CH_3\overset{OH}{\overset{|}{C}}{=}CH_2 + CH_2 = CH_2$$

$$\longrightarrow \text{cyclobutanol: } CH_2C(OH)\!-\!CH_2,\ CH_2\!-\!CH_2$$

$$CH_3CH_2\overset{O}{\overset{\|}{C}}CH_2CH_3 \xrightarrow[\text{process}]{\text{Type III}} CH_3CH_2\overset{O}{\overset{\|}{C}}\cdot\ \ \cdot CH_2\!-\!CH_2\!-\!H \longrightarrow CH_3CH_2\overset{O}{\overset{\|}{C}}H + CH_2 = CH_2$$

$$Ar\overset{O}{\overset{\|}{C}}Ar \xrightarrow[\text{photoreduction}]{R_2CHOH} Ar\overset{OH}{\overset{|}{\dot{C}}}Ar \longrightarrow Ar_2\overset{OH}{\overset{|}{C}}\!-\!\overset{OH}{\overset{|}{C}}Ar_2$$

cyclopentenone $\xrightarrow[\text{addition}]{\text{dimerization}}$ head-to-tail dimer + head-to-head dimer

4,4-diphenylcyclohexadienone $\xrightarrow{\text{dienone rearrangement}}$ 6,6-diphenylbicyclo[3.1.0]hexenone (Ph, Ph)

**Fig. 7.5** Photochemical reactions of carbonyl compounds.

reactivity of the singlet and triplet states is not well understood but the greater involvement of triplet states in chemical reactions is due in all probability to their longer lifetimes. As mentioned earlier singlet states decay within $10^{-8}$ seconds while the lifetimes of triplets are much longer and may vary from $10^{-4}$ seconds to several minutes.

To obtain a mechanistic picture of a photochemical reaction it is often useful to treat the excited molecule merely as a diradical but it must be borne in mind that many aspects of the excited state are as yet not well understood and have no analogies to usual ground state chemistry. The excited triplet state of carbonyl compounds may suffer bond homolysis, or take part in hydrogen abstraction reactions with the oxygen atom of the carbonyl moeity behaving like an alkoxy radical. These photochemical reactions of carbonyl compounds[4] may be classified as follows:

(1) Norrish type I process or $\alpha$-cleavage,
(2) Norrish type II process or $\beta$-cleavage,
(3) Norrish type III process,
(4) Photoreduction, and
(5) Additions and rearrangements,

and are illustrated in Fig. 7.5.

The photolysis of acetone gives methyl and acetyl radicals by $\alpha$-cleavage; subsequent reactions of these radicals then lead to a variety of products, as indicated below.

$$CH_3\overset{\overset{O}{\|}}{C}\text{—}CH_3 \rightarrow CH_3\overset{\overset{O}{\|}}{C}\cdot + \cdot CH_3$$

$$CH_3\overset{\overset{O}{\|}}{C}\cdot \rightarrow CH_3\cdot + CO$$

$$2CH_3\cdot \rightarrow C_2H_6$$

$$CH_3\overset{\overset{O}{\|}}{C}\cdot + CH_3\overset{\overset{O}{\|}}{C}CH_3 \rightarrow CH_3\overset{\overset{O}{\|}}{C}H + \cdot CH_2\overset{\overset{O}{\|}}{C}CH_3$$

$$CH_3\cdot + CH_3\overset{\overset{O}{\|}}{C}CH_3 \rightarrow CH_4 + \cdot CH_2\overset{\overset{O}{\|}}{C}CH_3$$

$$2CH_3\overset{\overset{O}{\|}}{C}\cdot \rightarrow CH_3\overset{\overset{O}{\|}}{C}\text{—}\overset{\overset{O}{\|}}{C}CH_3$$

$$2CH_3\overset{\overset{O}{\|}}{C}CH_2\cdot \rightarrow CH_3\overset{\overset{O}{\|}}{C}CH_2\text{—}CH_2\overset{\overset{O}{\|}}{C}CH_3$$

Photo-decarbonylations following $\alpha$ cleavage of cyclic ketones give non-stereospecific products. For example, *cis* or *trans* 2,6-dimethylcyclohexanone, (1*a*) or (1*b*), give the same mixture of *cis* and *trans* 1,2-dimethylcyclopentanes on decarbonylation and provide evidence for the diradical (2*a*) or (2*b*) being formed as intermediates.

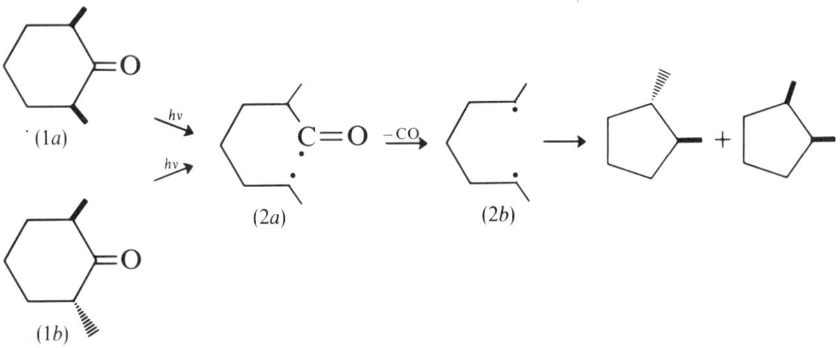

(1*a*) $\xrightarrow{h\nu}$ (2*a*) $\xrightarrow{-CO}$ (2*b*) $\longrightarrow$ + 

(1*b*) $\xrightarrow{h\nu}$

As indicated earlier ketones containing $\gamma$ hydrogen atoms (but not those without) can undergo two simultaneous photochemical reactions, one being Norrish type II or $\beta$ cleavage while the other is an intramolecular cyclobutanol formation. Intramolecular hydrogen transfers, like most $\gamma$ hydrogen abstractions, proceed via six-membered ring transition states and give diradical intermediates. Evidence for the formation of such intermediates is provided by the formation of a mixture of isomeric olefins (4*a*) and (4*b*) from the photolysis of either the *threo* ester (3*a*) or *erythro* ester (3*b*). A rapid carbon–carbon bond rotation in the diradical intermediate must have taken place before fragmentation. Further evidence for this intermediate has been obtained by photolysis of a ketone containing an optically active carbon centre at the $\gamma$ position, from which it was found that significant racemization had occurred in the cyclobutanol product.

Intermolecular hydrogen abstraction reactions by the photo-excited ketone molecule from hydrogen donating solvents are generally referred to as photoreductions. The formation of benzpinacol (5) in good yields from the irradiation of benzophenone in isopropanol was discovered by Ciamiciam and Silber as early as 1900. The high quantum yield ($\Phi \sim 2$) in this reaction indicates that hydrogen transfer from the isopropanol radical to benzophenone, as shown, is a very efficient step. Photoreduction appears general for a variety of aryl ketones and a number of solvents containing labile hydrogen can be used. Between these reactions of the excited triplet state carbonyl and those of alkoxy radicals many analogies can be drawn. A

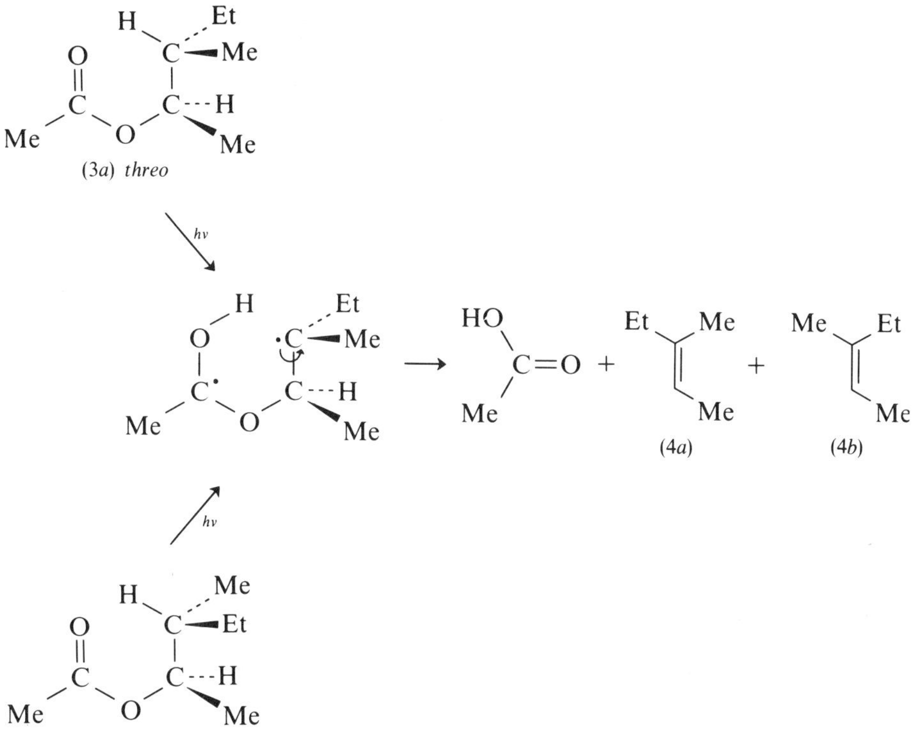

(3*a*) *threo*

$h\nu$

$\longrightarrow$ HO(Me)C=O + (4*a*) + (4*b*)

$h\nu$

(3*b*) *erythro*

comparison of the relative reactivity of benzophenone triplet to that of other radicals is given in Table 7.1 which indicates that the benzophenone triplet is more reactive (and hence less discriminating) than *t*-butoxy radicals in abstraction reactions. The polar character of the benzophenone triplet has recently been determined through study of the abstraction of hydrogen from nuclear substituted toluenes. Application of the Hammett equation

**Table 7.1** Relative Reactivity of C—H Bonds per Hydrogen Towards Various Radicals

| Type of C—H | $CH_3\cdot$ (182 °C) | Ph· (60 °C) | *t*-BuO· (40 °C) | Benzophenone Triplet (22 °C) | Cl· (23 °C) | F· (23 °C) |
|---|---|---|---|---|---|---|
| primary | 1 | 1 | 1 | 1 | 1 | 1 |
| secondary | 7 | 8–9.4 | 8–15 | 4.6 | 3.6 | 1.2 |
| tertiary | 50 | 44 | 44 | 9.9 | 4.2 | 1.4 |

$$\mathrm{Ph{-}\overset{\overset{\Large O}{\|}}{C}{-}Ph \xrightarrow{h\nu} Ph{-}\overset{\overset{\Large O^*}{\|}}{C}{-}Ph}$$

$$\mathrm{Ph{-}\overset{\overset{\Large O^*}{\|}}{C}{-}Ph + Me_2CHOH \longrightarrow Ph{-}\overset{\overset{\Large OH}{|}}{\underset{\bullet}{C}}{-}Ph + Me_2\dot{C}{-}OH}$$

$$\mathrm{Me_2\dot{C}{-}OH + Ph{-}\overset{\overset{\Large O}{\|}}{C}{-}Ph \longrightarrow Me_2CO + Ph\overset{\overset{\Large OH}{|}}{\underset{\bullet}{C}}Ph}$$

$$\mathrm{2Ph{-}\overset{\overset{\Large OH}{|}}{\underset{\bullet}{C}}{-}Ph \longrightarrow Ph_2\overset{\overset{\Large OH}{|}}{C}{-\!-}\overset{\overset{\Large OH}{|}}{C}Ph_2} \quad (5)$$

$$\mathrm{2Me_2\dot{C}OH \longrightarrow Me_2CO + Me_2CHOH}$$

gave a $\rho$ value of $-1.16$ at 22°C, correlation being made with $\sigma^+$ (Walling *et al.*)[5] showing the electrophilic character of the triplet species. On the other hand the Hammett plot for abstraction by the *t*-butoxy radical, correlated also with $\sigma^+$ gave $\rho$ as $-0.4$ only (at 40°C). A direct comparison of these $\rho$ values thus shows higher electrophilicity of the benzophenone triplet, provided of course that there is the same amount of bond stretching in the transition states.

Photochemical additions and rearrangements cover a large area of study and for further information the reader is referred to reference 1 and other related texts.

## 7.3 Photosensitized reactions

As discussed earlier photosensitized reactions can be effected by energy transfer from a species which absorbs light (sensitizer) to a substrate acceptor which then undergoes chemical change. The excitation energy transferred usually comes from the triplet state of the sensitizer since it is the triplet state which is longer lived, although singlet state energy transfer is also possible. A large number of photosensitized reactions involving olefins and dienes, themselves excellent triplet quenchers, have been studied. While intra- or intermolecular [2 + 2]-cycloadditions of olefins to give cyclobutanes may be brought about only after prolonged photolyses, in the presence of benzophenone as sensitizer these reactions can be effected rapidly and efficiently. Some of these reactions are depicted below.

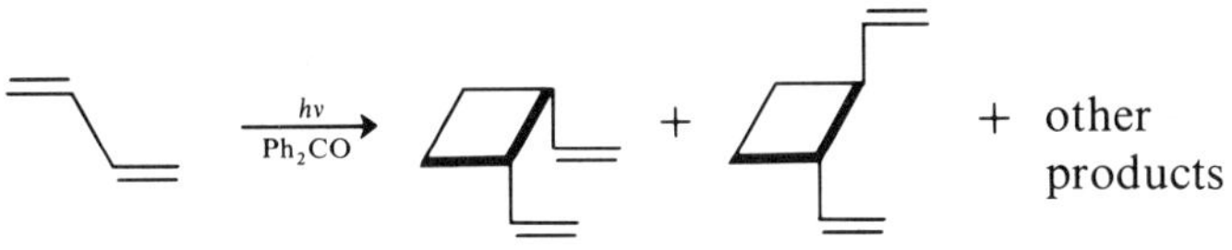

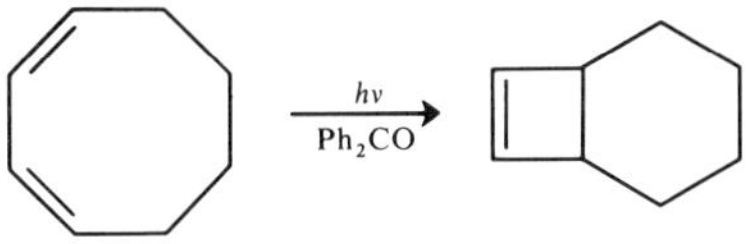

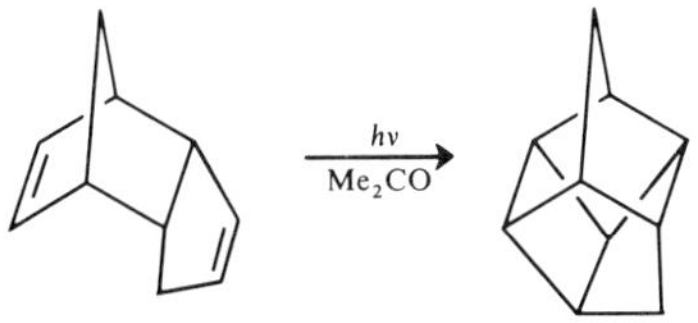

$\xrightarrow[Ph_2CO]{h\nu}$ + + other products

$\xrightarrow[Ph_2CO]{h\nu}$

$\xrightarrow[Me_2CO]{h\nu}$

Similarly the radical chain decarbonylation of aldehydes, which occurs with irradiation, is greatly accelerated through photosensitization with acetophenone.

Triplet-triplet energy transfer necessarily gives rise to an acceptor substrate which is in the triplet excited state. Use of this type of photosensitization is thus possible for the generation of triplet intermediates or radical pairs with two parallel spins. For instance, diazomethane and many diazoalkanes can be photolysed in the presence of benzophenone to give triplet state carbenes which, in contrast to the singlet state species, add non-stereospecifically to olefins (see also Chapter 17). The mechanism is given below in terms of triplet methylene, $CH_2\uparrow\uparrow$; however, the direct intervention of triplet diazomethane $(CH_2N_2)^T$ in the addition step cannot be ruled out.

$$CH_2N_2 + (Ph_2CO)^T \rightarrow (CH_2N_2)^T \rightarrow CH_2\uparrow\uparrow + N_2$$

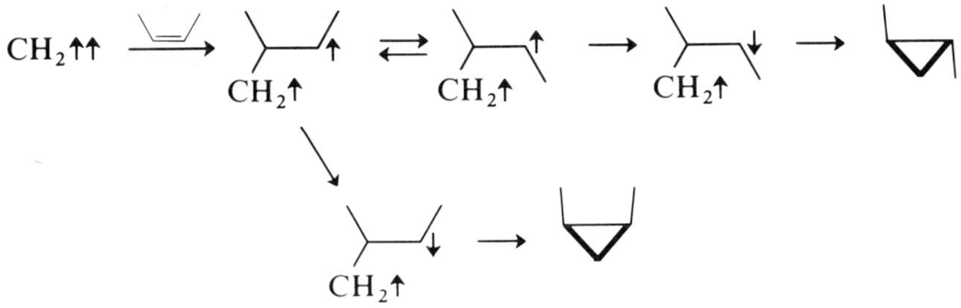

$CH_2\uparrow\uparrow$ ⟶ $CH_2\uparrow$ ⇄ $CH_2\uparrow$ ⟶ $CH_2\uparrow$ ⟶

$CH_2\uparrow$ ⟶

As stated earlier singlet energy transfer[6] is also known. The sensitized *cis-trans* isomerization of diphenylcyclopropane, formerly thought to be due to triplet-triplet energy transfer, is now established as due to singlet energy transfer. Conjugated dienes have been found to be good quenchers for singlet excited states of aromatic hydrocarbons (e.g., anthracene and naphthalene), but quenching or energy transfer in most cases do not lead to chemical reactions. Quadricyclene (6) however can isomerize to norbornadiene (7) by singlet-singlet energy transfer.

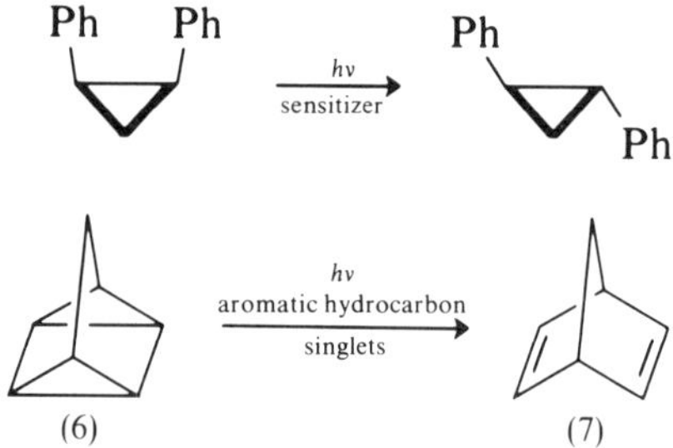

(6) (7)

The sensitized decomposition of open chain azocompounds usually proceeds by singlet energy transfer (Bartlett *et al.*)[7] but the cyclic azocompound (8) can undergo decomposition by triplet energy transfer to give mixtures of equal amounts of the *meso* and racemic cyclobutanes (9*a*) and (9*b*). Thermolysis or direct photolysis of the same azocompound (8) on the other hand gives the cyclobutanes (9*a*) in which there is substantial retention of the geometry of the azocompound precursor. This is explained by the formation in the triplet sensitized process of an intermediate triplet diradical (10) which having a relatively long lifetime undergoes rapid bond rotation before spin inversion and ring closure occur. It may be noted that any pair of radical centres with parallel (or unpaired) spins cannot form a bond (Pauli Principle); the spins must first be aligned anti-parallel or paired.

Me Et N=N Me Et — hv, triplet sensitizer → Me Et ↑ ↑ Me Et → Me Et Me Et + Et Me Me Et

(8) (10) (9*a*) (9*b*)

In general it has been observed that triplet diradicals (with parallel spins) produced by photolysis in the presence of a triplet sensitizer lead to non-stereospecific products, whereas direct thermolysis which generates singlet diradicals (*anti*-parallel spins) give products showing retention of the configuration of the starting materials.

1,2-Dioxetan (11) has been found to generate electronically excited states on thermal decomposition. The excitation energy can be transferred to a substrate acceptor causing it to undergo a chemical change as in photosensitization. It is noted that the thermal decomposition of this cyclic peroxide is the reverse of a [2 + 2]-cycloaddition reaction and as required by the conservation of orbital symmetry of a concerted reaction (see Section 3.7) allows the formation of an excited (luminescent) state in one of the carbonyl fragments.

This excited energy when transferred to *trans* stilbene brings about isomerization to *cis* stilbene and so essentially effects a 'photochemical reaction without light'.[8]

$$\underset{(11)}{\begin{matrix} O-O \\ | \quad | \\ >C-C< \end{matrix}} \xrightarrow{\Delta H} \begin{matrix} O\cdots O \\ \vdots \quad \vdots \\ >C\cdots C< \end{matrix} \longrightarrow \begin{matrix} O \\ \| \\ C \\ / \ \backslash \end{matrix} + \begin{matrix} O^* \\ \| \\ C \\ / \ \backslash \end{matrix} \text{ (excited state)}$$

## 7.4 Photodissociation–low temperature initiation

As has been noted photochemical cleavage provides a simple and useful method to generate free radicals for the initiation of radical chain reactions and polymerization. Moreover, by controlling the irradiation source a reaction can conveniently be initiated, regulated, or interrupted. The following equations exemplify some general methods of photochemical production of radicals, in each case absorption of light leading to a dissociative type of excited state is followed immediately by homolytic cleavage (refer to Fig. 7.2). When two different radicals are produced the chain carrier is underlined; chain sequences set up by these radicals through addition or abstraction reactions are treated in detail in other chapters.

$$X_2 \rightarrow 2X\cdot$$

$$RX \rightarrow \underline{R\cdot} + X\cdot \quad (X = Cl, Br)$$

$$ROX \rightarrow \underline{RO\cdot} + \underline{X\cdot}$$

$$HBr \rightarrow H\cdot + \underline{Br\cdot}$$

$$SO_2Cl_2 \rightarrow \underline{Cl\cdot} + \underline{\cdot SO_2Cl} \rightarrow SO_2 + Cl\cdot$$

$$RSH \rightarrow \underline{RS\cdot} + H\cdot$$

Peroxides RO—OR and peresters $RCO_2$—OR which usually absorb light appreciably in the region 340 nm to 220 nm can also be cleaved by direct photolysis, otherwise they may be decomposed by photosensitization. Thermolysis of these compounds will also give radicals but thermal processes suffer from the disadvantage that decomposition has to be carried out usually at a fixed range of relatively elevated temperatures. On the other hand, photochemical cleavage can be conducted at any convenient temperature and this is particularly useful when the reactants or products are unstable to heat. The photolysis of *t*-butyl peroxide solutions at low temperatures (down to $-100$ °C) has been used to generate short-lived radicals in solution for e.s.r. study where the more conventional flow methods are difficult to apply (Kochi and Krusic).[9] Azo compounds, well-known as polymerization initiators, are also extremely photo-reactive and generate radicals with the elimination of nitrogen. Diazo compounds on irradiation give carbenes. For instance, the photolysis of diazomethane in carbon tetrachloride first yields methylene which then initiates the reaction chain shown below to give pentaerythrityl tetrachloride (12). The reaction requires rapid 1,2 chlorine shifts in some of the intermediate radicals and is remarkable in that formation of the final product requires a total of no less than ten specific steps.

$$CH_2N_2 \xrightarrow{h\nu} :CH_2 + N_2$$

$$:CH_2 + CCl_4 \rightarrow \cdot CH_2Cl + \cdot CCl_3$$

$$Cl_3C\cdot + CH_2N_2 \rightarrow Cl_3CCH_2\cdot + N_2$$

$$CCl_3CH_2\cdot \rightarrow \cdot CCl_2CH_2Cl \xrightarrow{CH_2N_2} \cdot CH_2CCl_2CH_2Cl$$

$$\cdot CH_2CCl_2CH_2Cl \rightarrow ClCH_2\dot{C}ClCH_2Cl \rightarrow \rightarrow C(CH_2Cl)_4$$

(12)

It will be remembered that photolysis provides the only convenient method for the generation, and study by e.s.r., of very reactive triplet species, including carbenes and other diradicals. The photolysis of 3H-indazole (13) at liquid nitrogen temperature (77 K), for example, gives the triplet diradical (14*a*) which may be trapped by butadiene (a triplet scavenger) to give another triplet diradical (14*b*). The e.s.r. spectra of both of these diradicals have been recorded and identified by Closs and Kaplan.[10] The spectrum of the diradical tetramethylene-ethane (16), generated by the photodecarbonylation of (15) at 77 K, can similarly be obtained.

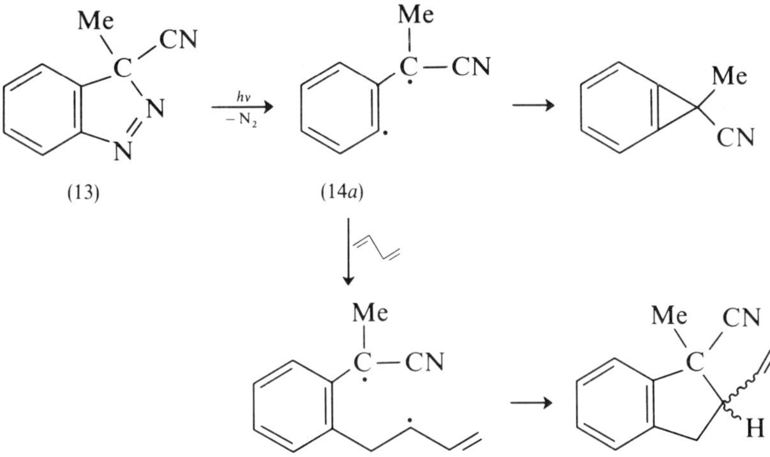

(13) (14a)

(14b)

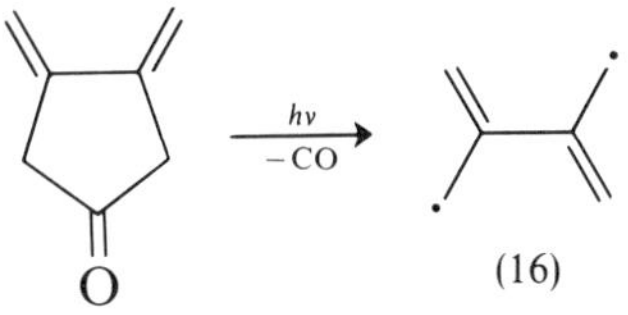

(15) (16)

## 7.5 The Barton reaction

The introduction of functional groups on to angular methyl groups in the steroid molecule (17), a key step in the synthesis of such important steroid hormones as aldosterone (18), had for many years presented serious difficulties. As the controlled reaction of the angular methyl group or its removal under mild conditions could not be achieved with the usual reagents, new methods of attacking had to be devised. The solution of this problem, a tribute to the ingenuity of Corey, Jeger, Barton, and others[11] was found in the generation of a reactive radical centre in close proximity to the methyl group to be attacked. The basic feature in these reactions lies in the fact that, in an intramolecular hydrogen abstraction process the six-membered transition state (19) involved affords specificity in abstraction of hydrogen atoms occurring four carbon atoms away from the radical centre, as illustrated in the reactions described below. Among these reactions one of the most useful is the photochemical decomposition of the alkyl nitrites (20) to the hydroxy oximes (21), known as the Barton reaction. This process, though

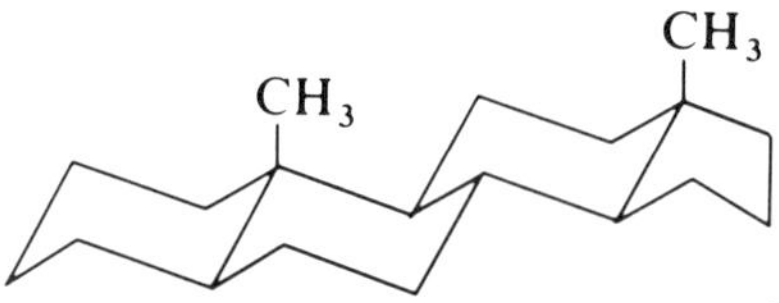

(17)

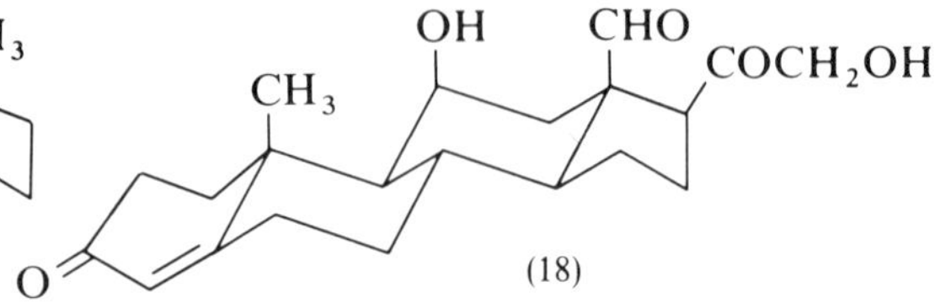

(18)

known for a long time, was left uninvestigated until Barton[12] realized its full synthetic potential. Its mechanism involves the formation of an alkoxy radical–nitric oxide radical pair in a solvent cage (22) followed by a rapid intramolecular hydrogen abstraction via a six-membered ring transition state (19) and subsequent recombination of the radical with nitric oxide.

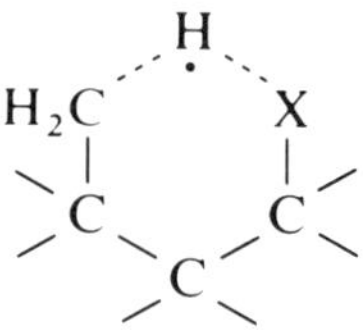

(19)

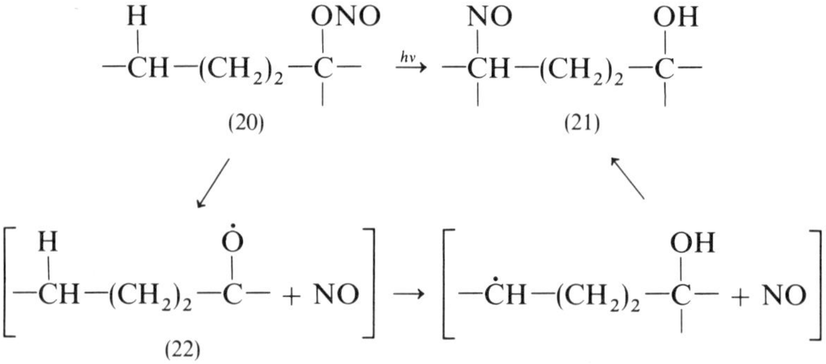

$$-\overset{\text{H}}{\overset{|}{\text{C}}}\text{H}-(\text{CH}_2)_2-\overset{\text{ONO}}{\overset{|}{\underset{|}{\text{C}}}}- \xrightarrow{h\nu} -\overset{\text{NO}}{\overset{|}{\underset{|}{\text{C}}}}\text{H}-(\text{CH}_2)_2-\overset{\text{OH}}{\overset{|}{\underset{|}{\text{C}}}}-$$

(20) (21)

$$\left[-\overset{\text{H}}{\overset{|}{\text{C}}}\text{H}-(\text{CH}_2)_2-\overset{\dot{\text{O}}}{\overset{|}{\text{C}}}- + \text{NO}\right] \rightarrow \left[-\dot{\text{C}}\text{H}-(\text{CH}_2)_2-\overset{\text{OH}}{\overset{|}{\underset{|}{\text{C}}}}- + \text{NO}\right]$$

(22)

The use of the Barton reaction for functionalization of steroidal angular methyl groups may be illustrated by the photolysis of (23) as sketched in the accompanying chart. It may be noted that in the conformationally rigid steroid system, the radical (24) may attack stereospecifically both angular methyl hydrogens which are in close proximity giving the radical (25) and the homo-allylic radical (26), the latter leading to formation of a cyclopropane ring.

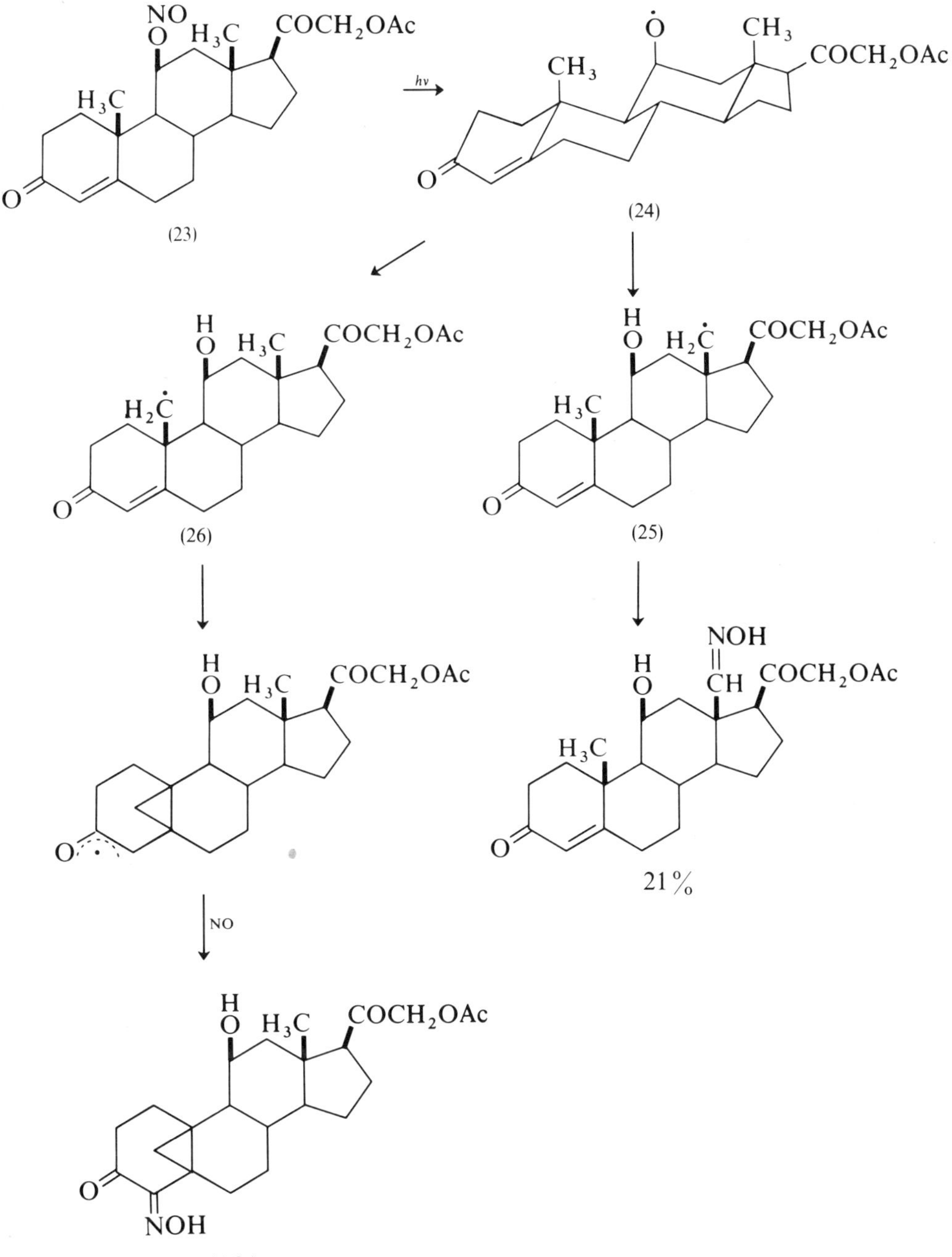

NO O $H_3C$ $COCH_2OAc$ $H_3C$ O

$h\nu$

(23)

$\dot{O}$ $CH_3$ $COCH_2OAc$ $CH_3$ O

(24)

H O $H_3C$ $COCH_2OAc$ $H_2\dot{C}$ O

(26)

H O $H_2\dot{C}$ $COCH_2OAc$ $H_3C$ O

(25)

H O $H_3C$ $COCH_2OAc$ O

NOH H O CH $COCH_2OAc$ $H_3C$ O

21%

NO

H O $H_3C$ $COCH_2OAc$ O NOH

45%

The abstraction of unactivated hydrogen atoms by alkoxy radicals, generated by photolytic or catalytic decomposition of hypohalities, is similar[11] (see Chapter 10).

## References

1 Several texts on Organic Photochemistry are available including the following:
*a* N. J. TURRO, *Molecular Photochemistry*, Benjamin, New York, 1967;
*b* D. C. NECKERS, *Mechanistic Organic Photochemistry*, Reinhold, New York, 1967;
*c* W. A. NOYES, G. S. HAMMOND, and J. N. PITTS, Eds. *Advances in Photochemistry*, Vols. I–III, Interscience, New York;
*d* R. O. KAN, *Organic Photochemistry*, McGraw-Hill, New York, 1966.
2*a* N. J. TURRO, *J. chem. Educ.* 1966, **43**, 13;
*b* G. S. HAMMOND in *Reactivity of the Photoexcited Organic Molecule*, Proceedings of the Thirteenth Conference on Chemistry at the University of Brussels, Interscience, London, 1967, p. 119.
3*a* E. M. KOSOWER, *An Introduction to Physical Organic Chemistry*, Wiley, 1968, p. 325;
*b* H. H. JAFFE and M. ORCHIN, *Theory and Applications of Ultraviolet Spectroscopy*, Wiley, New York, 1962.
4 N. C. YANG in *Reactivity of the Photoexcited Organic Molecule*, Interscience, London, 1967, p. 145.
5*a* C. WALLING and J. A. MCGUINESS, *J. Am. chem. Soc.*, 1969, **91**, 2053;
*b* C. WALLING and M. J. GIBIAN, *J. Am. chem. Soc.*, 1965, **87**, 3361.
6 S. L. MUROV, R. S. COLE, and G. S. HAMMOND, *J. Am. chem. Soc.*, 1968, **90**, 2957.
7 P. D. BARTLETT and N. A. PORTER, *J. Am. chem. Soc.*, 1968, **90**, 5317.
8*a* H. E. ZIMMERMAN, D. S. CRUMRINE, D. DÖPP, and P. S. HUYFFER, *J. Am. chem. Soc.*, 1969, **91**, 434;
*b* E. H. WHITE, J. WIECKO, and D. F. ROSWELL, *J. Am. chem. Soc.*, 1969, **91**, 5194.
9*a* J. F. KOCHI and P. J. KRUSIC, *J. Am. chem. Soc.*, 1968, **90**, 7156;
*b* *ibid.*, 1968, **90**, 7157.
10 G. L. CLOSS and L. R. KAPLAN, *J. Am. chem. Soc.*, 1969, **91**, 2168.
11 D. N. KIRK and M. P. HARTSHORN, *Steroid Reaction Mechanisms*, Elsevier, Amsterdam, 1968, Chapter 10.
12*a* D. H. R. BARTON, J. M. BEATON, L. E. GELLER, and M. M. PECHET, *J. Am. chem. Soc.*, 1960, **82**, 2640;
*b* D. H. R. BARTON and J. M. BEATON, *J. Am. chem. Soc.*, 1960, **82**, 2641.

# 8
# Redox Reactions

Redox reactions in general lead to the production of radicals and may be brought about by (a) electrolysis or (b) the action of alkali metals or transition metal ions on various organic compounds. These reactions represent a variety of oxidative as well as reductive processes.

Electrolysis provides a convenient method for the oxidation at the anode, or reduction at the cathode, of organic compounds. In addition to its many applications in synthesis it has, in affording a method for the continuous generation of radicals and radical ions, made possible numerous e.s.r. studies which could not otherwise have been carried out. Oxidation of organic compounds by certain metal ions in their high valency states, as well as reduction by metals or metal ions, have similarly found a place of importance in organic synthesis. The transition metals and their compounds, in particular, are widely used to generate radicals for polymerization processes and a host of other catalytic industrial processes.

## 8.1 Electrolysis

In an electrochemical reaction, which is a combination of a redox process and a chemical reaction, one might expect the electrode to be both a selective and a universal redox agent applicable for the mildest as well as the most vigorous conditions. In practice however, electrochemical processes are often accompanied by side reactions although these may be suppressed through control of current density, electrode potential or by variation of the composition of the electrode. As a convenient synthetic method one of the earliest instances is the Kolbe synthesis of simple alkanes, reported in 1849.

One-electron oxidation of a compound at the anode leads to radical cations while one-electron reduction at the cathode gives radical anions. Radical ions so produced are usually short-lived, unless they are considerably stabilized by resonance, and will therefore undergo further reaction with the solvent or with themselves. Thus the relatively stable *p*-substituted triphenylamine radical cation (1), a species isoelectronic to Gomberg's triphenylmethyl, can be generated by reversible one-electron transfer. The unsubstituted triphenylamine radical $Ph_3N^{\dot{+}}$, however, undergoes a rapid coupling reaction,[1] as shown.

$$(4\text{-MeOC}_6H_4)_3N \rightleftarrows (4\text{-MeOC}_6H_4)_3N^{\dot{+}} + e$$
$$(1)$$

$$Ph_3N^{\dot{+}} \longrightarrow Ph_2N\text{—}C_6H_4\text{—}C_6H_4\text{—}NPh_2 + H^+$$

In the Kolbe reaction, one of the most useful among electrochemical reactions,[2] oxidation of the carboxylate anion $RCOO^-$ takes place heterogeneously at the anode surface to give a radical which remains adsorbed on the electrode surface, shown below as (2). Decarboxylation of (2) gives the radical (3), which may dimerize or undergo side reactions, but usually takes the former course to give high yields of the dimer R—R. Various extensions of this reaction have been devised. The electrolysis of a mixture of two carboxylate anions, for example, or that of half esters of dicarboxylic acids, provides methods of synthesis for the unsymmetrical crossed Kolbe products (4) or diesters (5), respectively. Another interesting extension worth mentioning is the intramolecular coupling, via the diradical intermediate (6*a*) or the zwitterion (6*b*), in the conversion of the cyclobutane-1,3-dicarboxylic acid (6) to the bicyclic compound (7).

A large number of nitro and carbonyl compounds can readily be reduced at the cathode. Nitrobenzene for instance has long been known (Haber, 1900) to give, depending on the applied potential, current density and hydrogen ion concentration, predominantly one of a variety of products including nitrosobenzene, azoxybenzene, azobenzene or aniline. The reduction of the ketone

$$RCOO^- \xrightarrow{-e} \underset{(2)}{RCOO\cdot\,(\text{ads})} \xrightarrow{-CO_2} \underset{(3)}{R\cdot\,(\text{ads})} \longrightarrow R\text{—}R$$

(3) → abstraction and disproportionation

$$R\cdot\,(\text{ads}) \xrightarrow{-e} R^+$$

$$RCOO^- + R'COO^- \longrightarrow \underset{(4)}{R\text{—}R + R\text{—}R' + R'\text{—}R'}$$

$$RO_2C\text{—}(CH_2)_n\text{—}COO^- \longrightarrow \underset{(5)}{RO_2C\text{—}(CH_2)_{2n}\text{—}CO_2R}$$

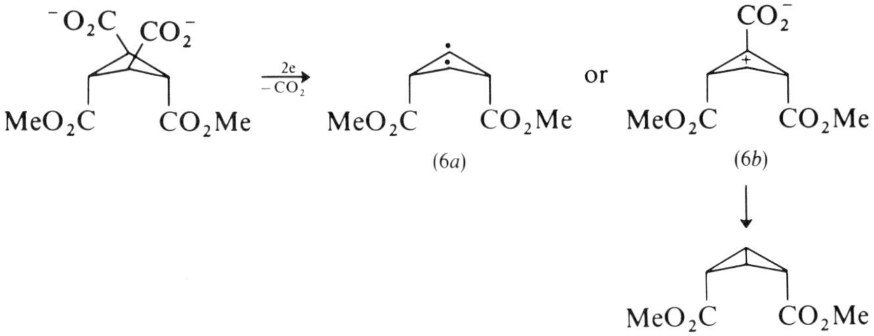

$$^{-}O_2C\ CO_2^- \text{(bicyclobutane)}\ MeO_2C\ \ CO_2Me \xrightarrow[-CO_2]{2e} MeO_2C\ \ CO_2Me\ (6a) \text{ or } MeO_2C\ \ CO_2Me\ (6b) \rightarrow MeO_2C\ \ CO_2Me\ (7)$$

(6*a*) (6*b*) (7)

$R_2CO$ gives the radical $R_2\dot{C}OH$ which dimerizes to the pinacol (8); the reduction of *p*-aminoacetophenone by this method affords a convenient synthetic route to the corresponding pinacol which cannot easily be obtained by the usual chemical methods.

$$R_2C{=}O \rightarrow R_2C{=}\overset{+}{O}H \xrightarrow{e} R_2\dot{C}{-}OH \rightarrow (R_2\overset{|}{C}{-}OH)_2$$

(8)

Aromatic systems may be reduced electrochemically: the cathodic reduction of naphthalene, for instance, gives a radical anion which can be trapped by carbon dioxide to give 1,4-dihydronaphthalene-1,4-dicarboxylic acid. Halides RX can be cleaved electrolytically, the ease of cleavage decreasing in the order I, Br, Cl. It is generally believed that reduction occurs by electron transfer to the halogen atom to give the radical R·, which is further reduced to the carbanion $R^-$. The electrolysis of $PhCMe_2{-}CH_2Cl$ gives predominantly *t*-butylbenzene and only 7% of the rearranged product, isobutylbenzene, indicating the intermediacy of the neophyl radical which, in contrast to the corresponding carbonium ion, is not prone to rearrangement (see Chapter 16). The reduction of vicinal dihalides gives olefins but electrolysis

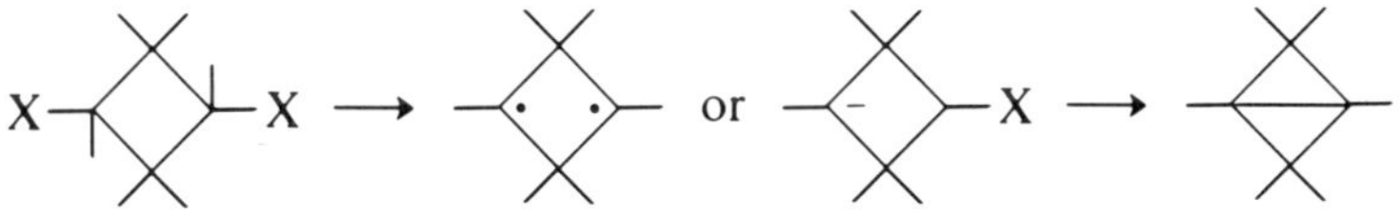

X–(cyclobutane)–X → (9*a*) or (9*b*)–X → (10)

(9*a*) (9*b*) (10)

Br–(cyclobutane)–$CH_2Br$ → (11)

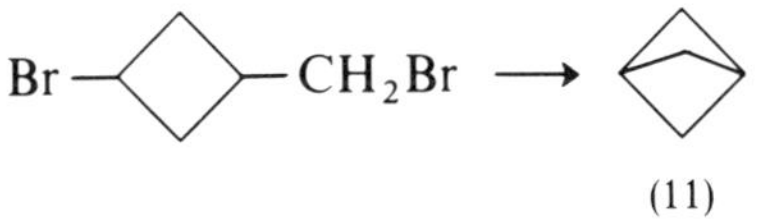

(11)

of 1,3-dibromopropane and 1,4-dibromobutane results in cyclization to cyclopropane and cyclobutane, respectively.[3] The latter reactions have been used for the syntheses of the bicyclo[1,1,0]butane derivatives (10) and bicyclo[1,1,1]pentane (11). The formation of (10) probably proceeds by way of the intermediates (9*a*) or (9*b*).

## 8.2 Redox reactions of transition metal ions

A large number of metal ions, particularly those of the transition metals, are capable of taking part in redox reactions.[4] In their low valency states they are strong reducing agents whereas in their high valency states they possess powerful oxidizing properties. Some ions can also function reversibly as oxidants and as reductants.

As mentioned in Chapter 5, free hydroxyl radicals[5] can be generated from ferrous ions and hydrogen peroxide (Fenton's reagent). This reagent can be used to oxidize almost all types of organic compounds; even benzene is oxidized to biphenyl and phenol. Free alkoxy radicals may also be produced by the reduction of peroxides or hydroperoxides with reducing metal ions (such as $Fe^{II}$, $Cr^{II}$, $Ti^{III}$, $Co^{II}$, or $Cu^{I}$) as depicted below:

$$Fe^{II} + HO\text{—}OH \rightarrow Fe^{III} + OH^- + HO\cdot$$

$$Ti^{III} + RO\text{—}OH \rightarrow Ti^{IV} + OH^- + RO\cdot$$

$$Cu^{I} + RCO_2\text{—}OR \rightarrow Cu^{II} + RCO_2^- + RO\cdot$$

$$Ag^{I} + {}^{-}O_3SO\text{—}OSO_3^- \rightarrow Ag^{II} + SO_4^{2-} + SO_4^{\dot{-}}$$

These reactions proceed at much lower temperatures than for the uninduced decomposition of the peroxides, and this is advantageous in polymerization initiation. The catalytic decomposition of peroxides and hydroperoxides by metal ions can also occur through oxidative pathways by metal ions such as $Fe^{III}$, $Co^{III}$, $Cu^{II}$, $Mn^{III}$, and $Ce^{IV}$. Hydroperoxides, for instance, readily generate peroxy radicals, as follows:

$$Fe^{III} + HO\text{—}O^- \rightarrow Fe^{II} + HO\text{—}O\cdot$$

$$Co^{III} + RO\text{—}OH \rightarrow Co^{II} + RO\text{—}O\cdot + H^+$$

The promotion of peroxide decomposition by metal ions has important implications. Autoxidations of rubber, plastic materials and lubricating oils, to mention a few, can be greatly accelerated, even in the dark, by the presence of such ions. Also, since the action of the metal ions depends on oxidation as well as reduction the catalyst would, through maintenance of an equilibrium such as $Fe^{IV} \rightleftarrows Fe^{V}$ or $Cu^{0} \rightleftarrows Cu^{I} \rightleftarrows Cu^{II}$, continually regenerate itself and thus retain its effectiveness. This equilibrium depends on the redox potential of the metal ion in its particular environment and certain additives such as complexing or chelating agents have been successfully employed to change the redox properties, resulting in the oxidation or the reduction reaction being considerably curtailed.

Homolytic oxidation of alcohols, aldehydes and phenols is also common. Fehling's and Benedict's solutions ($Cu^{II}$) and Tollen's reagent ($Ag[NH_3]_2^+$) are well known reagents for the oxidation of aldehydes; ferricyanide ($Fe^{III}[CN]_6^{3-}$) can be used to oxidize phenols, while alcohols are attacked by transition metal ions (or their complexes), e.g., $V^V$, $Mn^{III}$, $Ce^{IV}$, and $Co^{III}$, which have high redox potentials.

A number of synthetically important reactions in organic chemistry depend on the action of copper–copper ion catalysts.[6] The Sandmeyer reaction and related cyclization processes involving the aryldiazonium ion $ArN_2^+$ require such catalysts to promote the generation of the aryl radical Ar· or the carbonium ion $Ar^+$, as shown. The mechanism of the Pschorr cyclization reaction may be envisaged as proceeding initially by an electron transfer from the metal to the diazonium ion to give the radical (12) which, through an intramolecular addition reaction gives rise to the $\sigma$-complex radical (13) which in turn loses an electron to the metal ion thereby completing the cyclization process, as indicated. The special catalytic activity of copper and cuprous salts is attributed to their low oxidation potentials ($Cu \rightarrow Cu^I$, $E_0 = -0.13$ V; $Cu^I \rightarrow Cu^{II}$, $E_0 = 0.2$ V).

$$ArN_2^+ + Cu^0 \rightarrow Ar\cdot + N_2 + Cu^I$$

$$Ar\cdot + Cu^I \rightarrow Ar^+ + Cu^0$$

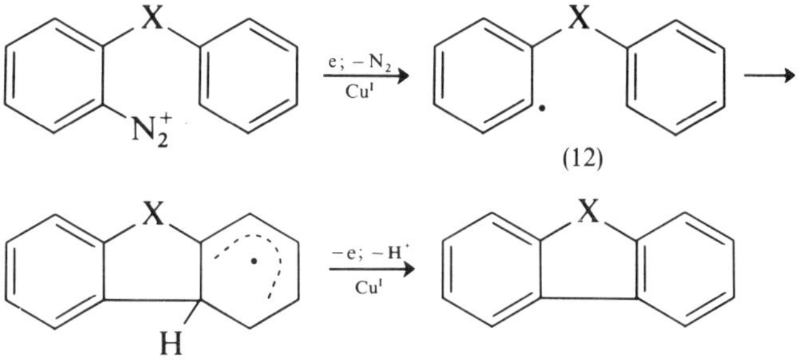

X, $N_2^+$, $\xrightarrow[Cu^I]{e;\ -N_2}$, X, (12), $\rightarrow$

X, H, (13), $\xrightarrow[Cu^I]{-e;\ -H^+}$, X

Two other synthetically useful reactions in which electron transfer is effected by copper ions may be mentioned. One is the decomposition of *t*-butyl peresters in the presence of substrates containing active hydrogen atoms (e.g., benzylic or allylic hydrogen atoms) and the other the reaction of diazonium salts with olefins carrying electronegative substituents. These are depicted below.

$$\mathrm{Bu^tO{-}O\overset{\displaystyle O}{\overset{\|}{C}}Ph + PhCH_2O\overset{\displaystyle O}{\overset{\|}{C}}CH_3 \xrightarrow{Cu^I} PhCHO\overset{\displaystyle O}{\overset{\|}{C}}CH_3 \; (\text{with } O{-}COPh \text{ on CH})} \qquad (50\%)$$

$$\mathrm{PhN_2^+Cl^- + CH_2{=}CHCN \xrightarrow{Cu^{II}} PhCH_2CHClCN} \qquad (81\%)$$

The reduction of organic halides by $Co^{II}$ and $Cr^{II}$ complex ions also deserves comment. Halpern and Maher[7] reported the formation of many stable organo-pentacyanocobaltate ions (14) by the reaction of pentacyanocobaltate ions ($Co^{II}[CN]_5^{3-}$) on organic halides. Similar reduction employing chromous ions[8] gives the organochromium species (15). One-electron reduction of carbonium ions by strongly reducing metal ions is also possible: for example, the triphenylmethyl carbonium ion is reduced by $V^{II}$ to the triphenylmethyl radical.

$$\mathrm{Co^{II}(CN)_5{}^{3-} + RX \rightarrow XCo^{III}(CN)_5{}^{3-} + R\cdot}$$

$$\mathrm{R\cdot + Co^{II}(CN)_5{}^{3-} \rightleftarrows \underset{(14)}{RCo^{III}(CN)_5{}^{3-}}}$$

$$\mathrm{Cr^{II} + RX \rightarrow XCr^{III} + R\cdot}$$

$$\mathrm{R\cdot + Cr^{II} \rightarrow \underset{(15)}{RCr^{II}}}$$

## 8.3 Reduction by dissolving metals

This type of reduction has found many applications in organic synthesis.[9] Strongly reducing metals such as the alkali metals are useful reductants but other metals like calcium, magnesium, zinc, tin, and iron, can be used as well. The reduction of organic compounds by metals can be regarded as an 'internal' electrolytic reduction, electron transfer from the metal generating radicals or radical anions. Treatment of benzophenone with sodium in an ethereal solution under an inert atmosphere of nitrogen, for example, gives the ketyl $Ph_2\dot{C}{-}O^-Na^+$ which exhibits a strong blue colour in solution and was probably the first radical anion to be discovered (Bechman, 1891). Similar reduction of polycyclic aromatic hydrocarbons, such as naphthalene and anthracene, yields relatively stable radical anions. An ethereal solution of sodium naphthylide, $C_{10}H_8^{\dot{-}}Na^+$, shows the properties of both carbanion and radical and can be employed as a strong reducing agent. With it the reduction of the optically active bromocyclopropane (16) was found to give a product with net retention of configuration. It is likely that an intermediate cyclopropyl radical (17*a*), which is configurationally labile (see Chapter 4), is formed but is rapidly trapped by reduction to the configurationally stable carbanion (17*b*) before complete racemization can occur.

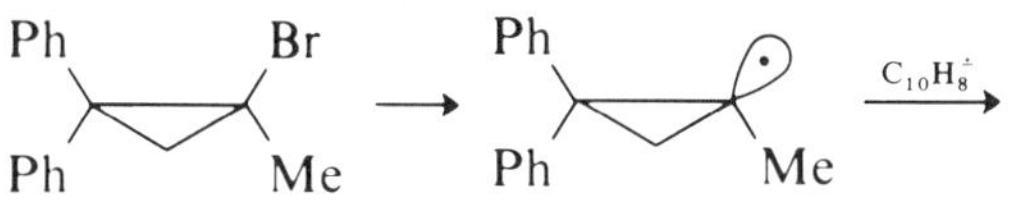

(16) (17*a*) 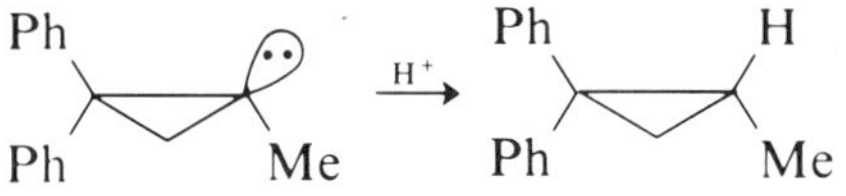 (17*b*)

Selective reduction of aromatic ring systems can be achieved by dissolving alkali metals in liquid ammonia (the Birch reduction).[10] This process involves solvated electrons, thus benzoic acid is reduced to the 1,4-dihydro derivative, probably via the intermediates (18*a*) and (18*b*). Formation of the radical ion (19) followed by coupling in all likelihood constitutes the reaction path in the bimolecular reduction of ketones to pinacols by magnesium amalgam. Similarly the reduction of nitrobenzene to aniline by a number of metals no doubt proceeds stepwise through a number of radical ions (20*a*–20*d*) as intermediates. Mention must also be made of the facile reduction of organic halides by alkali metals, the most familiar example being the Wurtz synthesis of hydrocarbons.

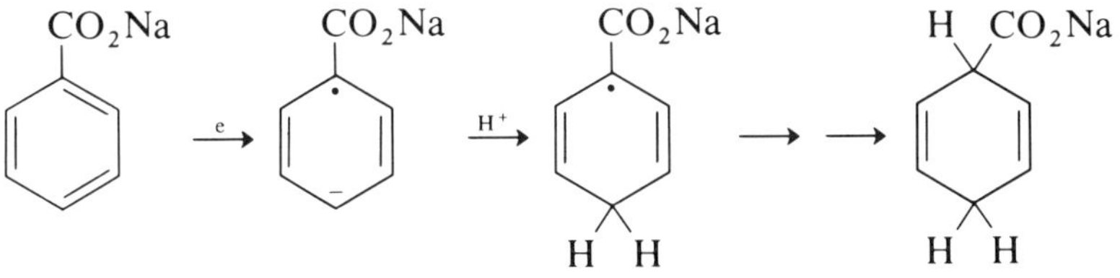

(18*a*) (18*b*)

$$Me_2C{=}O \xrightarrow{Mg} Me_2\dot{C}{-}O^-Mg^{2+} \rightarrow \begin{matrix} Me_2C{-}O^- \\ | \\ Me_2C{-}O^- \end{matrix} Mg^{2+}$$

(19)

$$Ph{-}\overset{+}{N}\begin{matrix} \diagup O^- \\ \diagdown O \end{matrix} \xrightarrow{e} Ph{-}\overset{\dot{+}}{N}\begin{matrix} \diagup O^- \\ \diagdown O^- \end{matrix} \xrightarrow{H^+} Ph{-}\overset{\dot{+}}{N}\begin{matrix} \diagup OH \\ \diagdown O^- \end{matrix} \xrightarrow{e}$$

(20*a*) (20*b*)

$$Ph{-}\ddot{N}\begin{matrix} \diagup OH \\ \diagdown O^- \end{matrix} \xrightarrow{H^+} Ph{-}\ddot{N}{=}O \xrightarrow{e} Ph{-}\underset{\cdot}{\ddot{N}}{-}O^- \xrightarrow{H^+}$$

(20*c*)

$$Ph{-}\underset{\cdot}{\ddot{N}}{-}OH \xrightarrow{e} Ph{-}\underset{\cdot\cdot}{\ddot{N}}{-}OH \xrightarrow{H^+} Ph{-}\underset{\displaystyle H}{\overset{\cdot\cdot}{\underset{|}{N}}}{-}OH \rightarrow \rightarrow PhNH_2$$

(20*d*)

## References


1 R. N. Adams, *Acc. Chem. Res.*, 1969, **2**, 175.

2 B. C. L. Weedon, *Adv. org. Chem.*, 1960, **1**, 1–34, R. A. Raphael, E. C. Taylor, and H. Wynberg, Eds., Interscience, New York.

3 J. D. Anderson, J. P. Petrovich, and M. M. Baizer, *Adv. org. Chem.*, 1969, **6**, 257–283, E. C. Taylor and H. Wynberg, Eds., Interscience, New York.

4 W. A. Waters, *Mechanisms of Oxidation of Organic Compounds*, Wiley, New York, 1964, Chapter 3.

5 W. A. Waters, *Organic Reaction Mechanisms*, Chemical Society Special Publication No. 19, London, 1965, p. 71.

6 R. A. Abramovitch, *Adv. Free-Radical Chem.*, 1967, **2**, 87–138, G. H. Williams, Ed., Academic Press, London.

7*a* J. Halpern and J. P. Maher, *J. Am. chem. Soc.*, 1965, **87**, 5361;

*b* J. Kwiatek, *Catal. Rev.*, 1967, **1**, 37.

8*a* F. A. L. Anet and E. LeBlanc, *J. Am. chem. Soc.*, 1957, **79**, 2649;

*b* W. C. Kray, Jr. and C. E. Castro, *ibid.*, 1964, **86**, 4603;

*c* J. K. Kochi and J. W. Powers, *J. Am. chem. Soc.*, 1970, **92**, 137;

*d* C. T. Loo, Lai-Yoong Goh, and S. H. Goh, *J.C.S. Dalton*, 1972, 585.

9 H. O. House, *Modern Synthetic Reactions*, Benjamin, New York, 1965, Chapter 3.

10*a* A. J. Birch, *Q. Rev. chem. Soc.*, 1950, **4**, 69.

*b* A. J. Birch and H. Smith, *ibid.*, 1958, **12**, 17.

*c* A. J. Birch and G. S. Rao, *Adv. org. Chem.*, 1972, **8**, 1, E. C. Taylor, Ed., Interscience, New York.

# 9

# Factors Governing Radical Reactions

Factors governing the course of radical reactions may be classified under stabilization, polar, steric, and solvent effects. Of these the first two have received by far the most intensive study. During the past decade investigations on the polar effect in particular have yielded many significant findings, following the availability of a method for its quantitative assessment. Studies on the steric factor, on the other hand, have remained qualitative in approach and are confined to a few instances where the consequences of its operation has become manifest. As to solvent effects observations of their playing a decisive role are relatively few and far between.

## 9.1 Stabilization versus polar effects

As was pointed out in Chapter 1, it was the recognition of the role of radicals as electrically neutral reacting entities in the erstwhile 'abnormal' reactions which afforded rationalization for the courses these reactions take. Being electrically neutral, radicals in their attack on other bodies, unlike ions, were originally expected to be indifferent to the electron density at the site of attack and therefore free from control by polar effects. It would therefore follow that the only factor directing the course of a radical reaction, provided steric and solvent effects do not intervene, would be the relative stability of the reaction intermediate or more exactly, the incipient radical in the transition state. This early generalization still holds with reactions in which the attacking radicals are alkyl radicals, and with reactions requiring small activation energies, such as addition reactions.

To illustrate how this generalization has been applied to homolytic reactions one may take the classical example of the addition of hydrogen bromide to an olefin with a terminal double bond. This takes place initially by the addition of the bromine atom to the terminal carbon atom because the radical (1*a*) so formed is more stabilized, through hyperconjugation, than the alternative radical (1*b*), and hence there is preference for the formation of (1*a*) over that of (1*b*). The attack of the phenyl radical on the *o* and *p* positions of nitrobenzene, which at one time was a puzzling phenomenon, can similarly be explained in terms of the greater stability, through delocalization of the unpaired electron, of the radical intermediates (2*a*) and (2*b*) involving initial addition at the *o* and *p* positions, relative to attack at the *m* position. The abstraction of hydrogen atoms from ethylbenzene, say by the methyl radical, would be expected to occur at the $\alpha$, rather than at the $\beta$ position, as is indeed the case, because the radical so generated is relatively more stabilized through resonance afforded by the benzene ring. (In the knowledge we now have of the nature of the transition state, however, it must be pointed out that the stabilization effect can accelerate the formation of radicals only when the transition state occurs late in the reaction coordinate and so already has strong radical character. For example, a comparison of the relative rates of hydrogen abstraction from propane and toluene by various radicals (see Table 3.6 in Chapter 3) shows that the effect on rate due to stabilization by the benzene ring is pronounced only in abstraction by the bromine atom in which there is significant bond breakage in the transition state.)

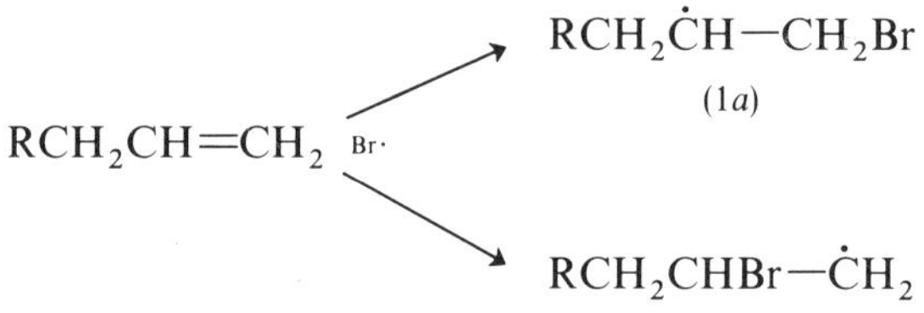

$$RCH_2CH{=}CH_2 \xrightarrow{Br\cdot} RCH_2\dot{C}H{-}CH_2Br \quad (1a)$$

$$RCH_2CH{=}CH_2 \xrightarrow{Br\cdot} RCH_2CHBr{-}\dot{C}H_2 \quad (1b)$$

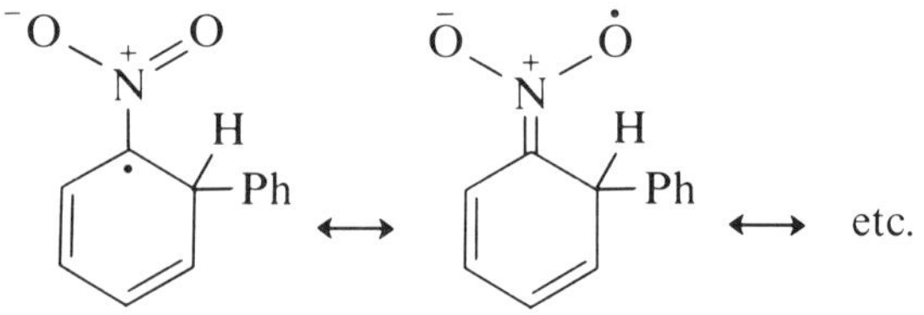

(2*a*)

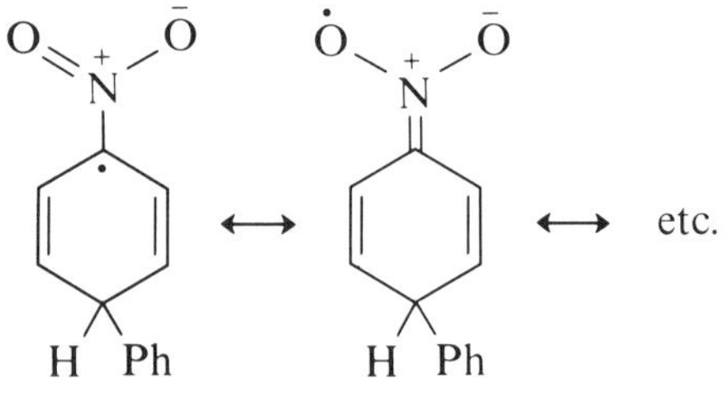

(2*b*)

In the last case quoted above, the benzene ring directs the course of reaction through its stabilizing influence on the intermediate radical. Other groups such as carbonyl, ester, cyano, or even methyl which can similarly delocalize the unpaired electron will also stabilize radical centres to which they are attached and when more than one such group are present in a molecule uncertainty arises as to which group will, through its superior stabilizing power, dictate the course of the reaction. For example, in the addition of the radical R· to the olefinic bond in the ketonic ester $CH_3COCH{=}CHCOOMe$, will it be the radical $CH_3COCHR{-}\dot{C}HCOOMe$ or the isomeric $CH_3CO\dot{C}H{-}CHRCOOMe$ through which addition will proceed? Or take the abstraction of hydrogen from hydrocinnamic acid, $PhCH_2CH_2COOH$, will it be the hydrogen atom on carbon 2 or that on carbon 3 which suffers abstraction? The course these reactions take might be predicted with some confidence if the relative stabilizing effects of the ketonic and ester groups in the first case, and that of the benzene ring and the carboxyl group in the second, are known. Thus a knowledge of the relative stabilizing capacities of substituent groups would be very useful in explaining and even predicting the course a reaction takes.

An approach to this problem has been made available through an intramolecular competition addition reaction to the unsymmetrically substituted olefin, $XCH{=}CHY$, in which X and Y are substituents whose stabilizing capacities are to be compared. Addition of R· to this olefin would give $X\dot{C}H{-}CHRY$ or $XCHR{-}\dot{C}HY$ according to whether X or Y, respectively, stabilizes the intermediate alkyl radical to the greater extent. A systematic study of the addition of bromotrichloromethane and aliphatic aldehydes to a series of olefins (X, Y = Ph, CO, COOMe etc.) yielded a scale of relative stabilizing capacities as follows (Huang):[1]

$$\mathrm{Ph > CO \sim CN > COOMe \sim COOH > CH_3 > H}$$

While obviously limited to reactions in which the stabilization effect plays a dominant role, this scale has found useful application in a number of radical reactions. (For a more detailed description of the method the reader is referred to Section 12.5.)

The ability of a substituent group to delocalize the odd electron of a radical can now be studied directly by electron spin resonance. For example by observing the decrease of spin density at the methyl group of the radical centre $Me\dot{C}(X){-}$ this ability of the substituent X has been found to be approximately of the order:[2]

$$\mathrm{EtO{-} > Et\overset{\overset{\displaystyle O}{\|}}{C}{-} > HO{-} > CN > Me > HOCH_2{-} > RO\overset{\overset{\displaystyle O}{\|}}{C}{-}}$$

Although the thesis that the course of a homolytic reaction was controlled solely by the relative stability of the radical intermediate seemed to hold for most radical reactions encountered, evidence was soon found to show that to ignore all participation of effects of a polar nature was an oversimplification. Actually, among themselves radicals spread over a range of polarity, even though compared with positive and negative ions they are relatively neutral entities. Thus alkyl and benzyl radicals may be classified as nucleophilic or donor radicals; their weakly nucleophilic character is not unexpected considering the greater ease of formation of carbonium ions than carbanions. The majority of other radicals, such as atomic halogens, alkoxy and peroxy radicals, however, have electrophilic or acceptor properties,[3] since the unpaired electron is situated on electronegative atoms, and such properties vary from radical to radical. Where the attacking radical has a relatively pronounced polar character, it would not be surprising if a polarized transition state is called into play. Take for example the typical reaction of hydrogen abstraction by the radical X· from toluene: whereas the transition state may adequately be represented by (3*a*) when the abstracting radical is the relatively neutral methyl, polar structures such as (3*b*) may well contribute significantly, and indeed do so, when abstraction is effected by the electrophilic chlorine atom. In this suggestion it is also evident that nuclear substituents which favour a polar transition state will facilitate the reaction going through such a transition state.

$$\underset{(3a)}{\mathrm{PhCH_2\cdot\ \ \dot{H}\ \ \dot{X}}} \longleftrightarrow \underset{(3b)}{\mathrm{Ph\overset{+}{C}H_2\ \ \dot{H}\ \ \bar{X}}}$$

The extent to which polar effects assert themselves, when they do so, is understandably much smaller in magnitude than in ionic reactions, in which charged particles participate. Nevertheless, in a number of cases, such effects have been known to completely change the course of the reaction, and in a few others, they apparently make all the difference of the reaction occurring or not occurring at all.

Among early examples of the polar effect may be quoted Kharasch's finding[4] in the abstraction of hydrogen atoms from 1-ethyl-2,3,4,5,6-pentachlorobenzene (4), from which atomic chlorine removes a $\beta$ hydrogen atom while bromine removes an $\alpha$ hydrogen atom. These results are best explained as follows. When the attacking entity is chlorine the overriding repulsive polar effect between the electrophilic chlorine atom and the nuclear chlorine atoms causes reaction to occur at the $\beta$ carbon. With atomic bromine, however, resonance stabilization offered by the benzene ring predominates over the weaker polar repulsive forces.

$$C_6Cl_5CH_2CH_3 \text{ (4)} \xrightarrow{Br\cdot} C_6Cl_5CHBrCH_3 \quad (100\%)$$

$$C_6Cl_5CH_2CH_3 \text{ (4)} \xrightarrow{Cl\cdot} \underset{(40\%)}{C_6Cl_5CHClCH_3} + \underset{(60\%)}{C_6Cl_5CH_2CH_2Cl}$$

Me· → Cl–C₆H₄–CH₂—O—CH₂–C₆H₅ ← Buᵗ O· and ArCOOO·

(5)

In the cleavage of *m*- and *p*-chlorodibenzyl ether (5), alkyl radicals attack mainly the benzylic hydrogen atom next to the substituted ring, whereas *t*-butoxy and benzoyloxy radicals abstract hydrogen from the other benzylic carbon atom.[5] It appears that the former carbon atom has been rendered electron-poor by the chloro substituent, and thus the electrophilic *t*-butoxy and benzoyloxy radicals avoid this position.

The strongly electrophilic nature of atomic chlorine is demonstrated by its different rates of attack on ethane, methane and halogenated methanes (Table 9.1). The introduction as substituent of chlorine atoms into methane causes an almost uniform decrease in the activation energy for attack by methyl radicals, but an increase for attack by atomic chlorine.[6]

**Table 9.1** Hydrogen Abstraction from Ethane, Methane and Chlorinated Methanes by Methyl Radicals and Atomic Chlorine*

| | Cl· | | Me· | |
|---|---|---|---|---|
| | log *A* | $E_\alpha$ | log *A* | $E_\alpha$ |
| $CH_4$ | 10.4 | 3.9 | 11.6 | 12.8 |
| $C_2H_6$ | 11.1 | 1.0 | 11.3 | 10.4 |
| $CH_3Cl$ | 10.7 | 3.4 | 11.8 | 9.4 |
| $CH_2Cl_2$ | 11.6 | 5.5 | 11.3 | 7.2 |
| $CHCl_3$ | 11.6 | 6.5 | 10.8 | 5.8 |

* Data from ref. 6; *A* in $mole^{-1}$ l. $s^{-1}$ and $E_\alpha$ in kcal $mole^{-1}$.

Polar effects have also been demonstrated in the chlorination of several derivatives of butane.[6,7] The results given in Table 9.2 show that the α position and the β position to a lesser extent, are deactivated by electron withdrawing groups while the γ position remains apparently unaffected. This is explicable in terms of an inductive polar effect, the influence of which

**Table 9.2** Relative Reactivities of Hydrogen Atoms in Derivatives of Butane towards Radicals*

| Radical | $T$, °C | X— | Relative Reactivities† —$CH_2$— | —$CH_2$— | —$CH_2$— | —$CH_3$ |
|---|---|---|---|---|---|---|
| Cl· | 35° | H— | 1 | 3.95 | 3.95 | (1) |
| F· | 20° | F— | 0.3 | 0.8 | 1.0 | (1) |
| Cl· | 35° | F— | 0.8 | 1.6 | 3.7 | (1) |
| Br· | 146° | F— | 10 | 9 | 88 | (1) |
| Cl· | 35° | Cl— | 0.7 | 2.2 | 4.2 | (1) |
| Br· | 160° | Cl— | 34 | 32 | 80 | (1) |
| Cl· | 75° | $F_3C$— | 0.04 | 1.2 | 4.3 | (1) |
| Br· | 160° | $F_3C$— | 1.0 | 7.0 | 90 | (1) |
| Cl· | 60° | F—C(=O)— | 0.08 | 1.5 | 4.2 | (1) |
| Cl· | 75° | Cl—C(=O)— | 0.2 | 2.1 | 3.9 | (1) |
| Cl· | 100° | Me—C(=O)O— | 0.1 | 2.1 | 4.0 | (1) |
| Cl· | 75° | MeOC(=O)— | 0.4 | 2.4 | 3.6 | (1) |
| Br· | 160° | MeOC(=O)— | 41 | 35 | 77 | (1) |
| Cl· | 100° | $CF_3$C(=O)O— | 0.2 | 1.3 | 3.6 | (1) |
| Br· | 150° | $CF_3$C(=O)O— | 2 | 7 | 66 | (1) |

* Data from ref. 6 and 7.
† Reactivity per hydrogen relative to the primary δ hydrogen.

decreases rapidly with distance. The stabilization ability of various substituent groups (e.g., Cl and C=O) is seen to be unimportant here in this low activation energy reaction, and polar effects apparently predominate over bond energy considerations. Data obtained for bromination also demonstrate the deactivation of the α and β positions by the substituent, but the effect is less pronounced than that in chlorination. As would be anticipated for this high activation energy reaction, polar deactivation of the α position is compensated to some extent through resonance stabilization, for example by the carbonyl group.

A recent observation made by Huang, Lee, and Ong[8] illustrates the extreme case where polar effects appear responsible for the occurrence, or otherwise, of a reaction. In the reaction of toluene or of benzyl methyl ether with *t*-butoxy radicals (liberated from the peroxide) in carbon tetrachloride as the solvent, it has been found that, whereas benzyl radicals derived from the former substrate fail to abstract chlorine atoms from carbon tetrachloride, α-methoxybenzyl radicals generated from the ether are able to do so thereby liberating trichloromethyl radicals and initiating a chain reaction as shown. It is believed that in the α-methoxybenzyl radical the presence of the α-methoxy group makes possible the participation of polar transition states such as (6) which facilitate the reaction.

$$\mathrm{PhCH_2OMe + Bu^tO\cdot \rightarrow Ph\dot{C}HOMe + Bu^tOH}$$

$$\mathrm{Ph\dot{C}HOMe + CCl_4 \rightarrow \left[\begin{array}{l}\mathrm{PhCH\cdots\dot{C}l\cdots \underset{\delta^-}{CCl_3}} \\ \quad\vdots \\ \quad \underset{\delta^+}{\mathrm{OMe}}\end{array}\right] \rightarrow \begin{array}{l}\mathrm{PhCHCl} \\ \quad | \\ \quad\mathrm{OMe}\end{array} + \cdot CCl_3}$$

(6)

$$\mathrm{\cdot CCl_3 + PhCH_2OMe \rightarrow HCCl_3 + Ph\dot{C}HOMe}$$

In copolymerization processes the regularity of alternation between different monomer units along polymer chains, referred to as the alternating effect,[9] is due to polar effects. For example, in the copolymerization of styrene and maleic anhydride, polarized transition states such as (7*a*) and (7*b*) must be favourable due to a suitable combination of an electron donor (styrene) and an electron acceptor (maleic anhydride). In cross-termination the polarized transition state (7*c*) is also favoured for the same reason. In general it has been found that when two types of radicals are present in a system, cross-coupling plays a more important role than the dimerization of individual radicals.

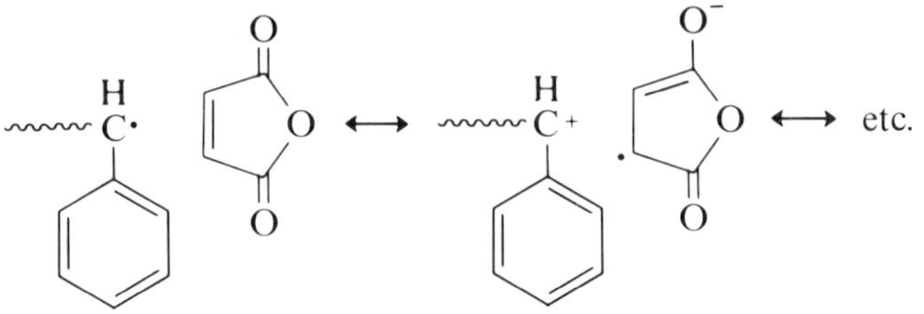

(7*a*)

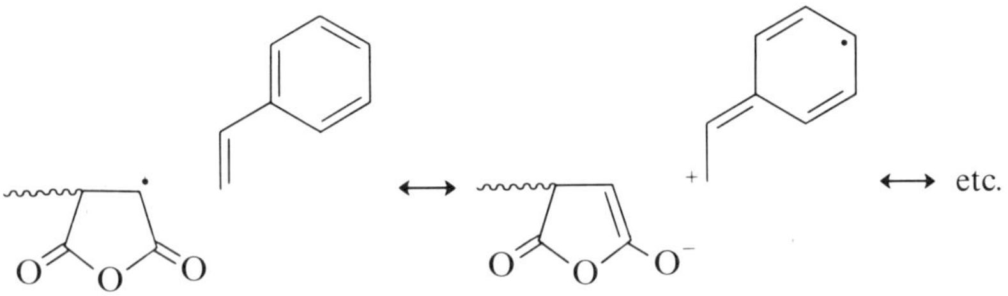

(7*b*)

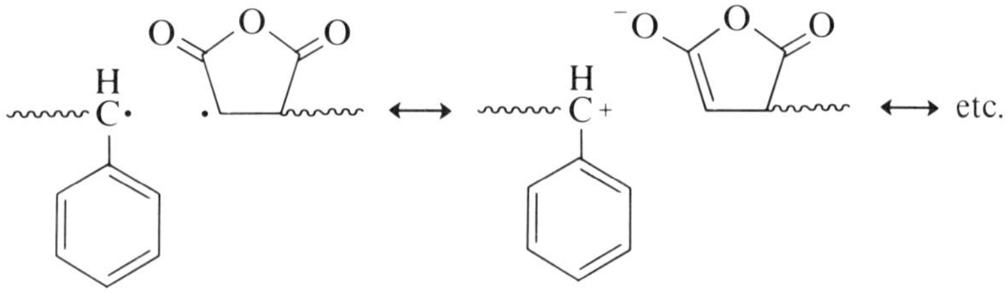

(7*c*)

## 9.2 Evaluation of polar effects by the Hammett relationship

By far the most effective approach to the study of polar effects has been through application of the Hammett equation, $\log k/k_0 = \rho\sigma$. As explained in Section 3.3, the magnitude of the reaction constant, given by the slope $\rho$ of the plot, gives a measure of the extent to which polar effects operate. This method has been put to good use in a large number of aromatic systems, mostly derivatives of toluene, following the work in the early 1950's by Kooyman and by Walling,[10] and their associates, who were pioneers in this area of investigation.

Subsequent work has shown that for the majority of systems studied, $\log k/k_0$ is in fact better correlated with the $\sigma^+$ constants of Brown and Okamoto (see Section 3.3) which are derived from the $S_N1$ solvolysis of *t*-cumyl chlorides involving the charged transition state (8*a*). Such a correlation in a hydrogen abstraction reaction would therefore imply a charged transition state (8*b*) similar to (8*a*) and it follows that the effect of a nuclear substituent on the reactivity of the side chain towards the abstracting radical is dependent on the extent to which the substituent stabilizes this polarized transition state.

$$\overset{\delta+}{\mathrm{Ar}}\text{:--:}\underset{\mathrm{Me}}{\overset{\mathrm{Me}}{\mathrm{C}}}\cdots\overset{\delta-}{\mathrm{Cl}}$$

(8*a*)

$$\overset{\delta+}{\mathrm{Ar}}\text{:--:}\mathrm{C}\cdots\dot{\mathrm{H}}\cdots\overset{\delta-}{\mathrm{X}}$$

(8*b*)

As an example may be cited the abstraction of benzylic hydrogen atoms from substituted toluenes by atomic chlorine.[10b] Plotting log $k/k_0$ against $\sigma^+$ as shown in Fig. 9.1, a straight line with a slope of $-0.66$ is obtained. The negative sign in $\rho$ here indicates the electrophilic character of the attacking radical, atomic chlorine, and the value of $\rho$ gives the magnitude of the polar effect.[25]

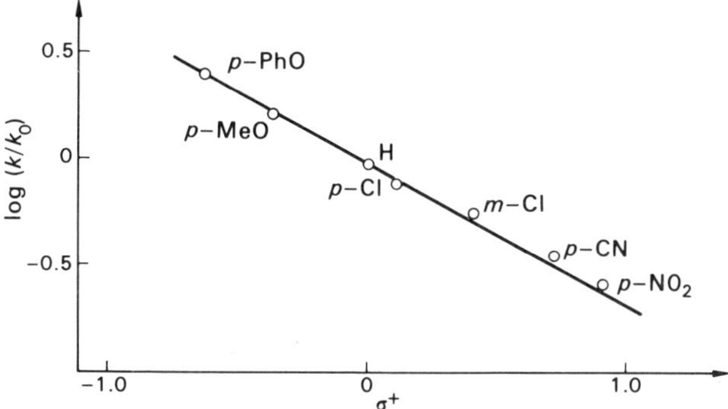


**Fig. 9.1** Hydrogen abstraction by atomic chlorine from substituted toluenes in carbon tetrachloride at 70 °C.[10b]

The results of Hammett studies on hydrogen abstraction from toluene by a large variety of radicals are collected in Table 9.3. As can readily be seen reasonable correlation may be obtained using either $\sigma$ or $\sigma^+$ constants, the choice of which is, however, not predictable as this depends on the nature of the transition state. In accordance with the nature of these substituent constants as defined, $\sigma$ constants would be expected to be generally useful

**Table 9.3** Abstraction of Benzylic Hydrogen Atoms from Nuclear Substituted Toluenes*

| Radical (source) | Solvent ($T$, °C) | $\rho\sigma$ or $\rho\sigma^+$ |
|---|---|---|
| Br· ($Br_2$, fast)† | $C_6H_6$ (80°) | $-1.36\sigma^+$ |
| Br· ($Br_2$, slow)† | $C_6H_6$ (80°) | $-1.07\sigma^+$ |
| Br· (NBS)‡ | $C_6H_6$ (80°) | $-1.46\sigma^+$ |
| Br· (NBTFS)‡ | $C_6H_6$ (80°) | $-1.45\sigma^+$ |
| Br· (NBTMS)‡ | $C_6H_6$ (80°) | $-1.36\sigma^+$ |
| Br· ($Br_2$, fast)† | $C_6H_6$ (19°) | $-1.76\sigma^+$ |
| Br· (NBTMS) | $C_6H_6$ (19°) | $-1.76\sigma^+$ |
| Br· (NBS) | $CCl_4$ (80°) | $-1.38\sigma^+$ |
| Cl· ($Cl_2$) | $CCl_4$ (40°) | $-0.66\sigma^+$ |
| $Bu^tO$· ($Bu^tOC_2O_4OBu^t$) | MeCN (40°) | $-0.39\sigma^+$ |
| $Bu^tO$· ($Bu^tOC_2O_4OBu^t$) | PhCl (40°) | $-0.32\sigma^+$ |
| $Bu^tO$· ($Bu^t_2O_2$) | Reactants + $\gamma$-picoline (110°) | $-0.646\sigma^+$ |
| $Bu^tO$· ($Bu^tOCl$) | Reactants (40°) | $-0.82\sigma$ |
| $Bu^tO$· ($Bu^tOCl$) | $CCl_4$ (40°) | $-0.68\sigma^+$ |
| $Ph_2\dot{C}—\dot{O}$ ($Ph_2CO$, $h\nu$) | reactants + $C_6H_6$ (22°) | $-1.16\sigma^+$ |
| $Cl_3C$· ($BrCCl_3$) | PhCl (50°) | $-1.46\sigma^+$ |
| $ClO_2S$· ($SO_2Cl_2$) | $CCl_4$ (40°) | $-0.56\sigma$ |
| ROO· (RH, $O_2$) | reactants (90°) | $-0.7\sigma$ |
| $p$-$NO_2C_6H_4$· ($ArN_2CPh_3$) | reactants (60°) | $-0.59\sigma$ |
| Ph· ($PhN_2CPh_3$) | reactants (60°) | $-0.1\sigma$ |
| $p$-$MeC_6H_4$· ($ArN_2CPh_3$) | $CCl_4$ (60°) | $-0.1\sigma$ |
| Me· (MeHgI, $h\nu$) | reactants (100°) | 0 |

* From ref. 12, 13, and 15.
† Fast refers to presence of excess molecular bromine.
‡ NBS, NBTFS, and NBTMS are $N$-bromosuccinimide, $N$-bromo-tetrafluorosuccinimide, and $N$-bromo-tetramethylsuccinimide respectively.

when the substituents assert their influences primarily through electrostatic interactions (inductive and field effects). This means that reactivity is determined mainly by the electron density of the C—H bond undergoing abstraction. On the other hand good correlations with $\sigma^+$ constants would imply that resonance interactions are important in the transition state and that there is considerable electron deficiency at the reaction site.[11] This is not unexpected since $\sigma^+$ constants incorporate a resonance factor for the stabilization of carbonium ion-like transition states. In fact the distinction between correlation with $\sigma$ or with $\sigma^+$ lies in the use of substituents such as the $p$-methoxy group which can strongly stabilize a carbonium ion structure.

From the data in Table 9.3 it can also be seen that the value of the reaction constant is very sensitive to the conditions of the reaction. Thus the $\rho$ values for the bromination of toluene with bromine have been found by Pearson and Martin[12] to depend on whether the bromine is added rapidly (i.e., in excess) or slowly. In the latter case it was found that hydrogen bromide, a reaction product, participates in the radical chain reaction making the results unreliable. Similarly the use of *t*-butyl hypochlorite as a source of *t*-butoxy radicals can give misleading results since chains involving chlorine atoms have also been detected in the reaction.[13] Where solvents are used care must be taken to ascertain that the added solvent does not participate in the radical chains.

The Hammett equation has been applied to a large number of other abstraction reactions involving various aromatic systems and a variety of abstracting radicals. Some of the more significant results are summarized in Table 9.4. From these studies it has become possible to draw a number of conclusions bearing on the operation of polar influences in radical reactions. These are given below.

### (a) *Polarity of the attacking radical*

For a substance entering into reaction with different radicals under a given set of conditions the extent to which polar effects are called into play is dependent on the polarity of the attacking radical. Although this condition might have been taken for granted, there has in fact been, until recently,

**Table 9.4** Hammett Correlations of Various Radical Reactions*

| Radical (source) | Substrate† | Solvent ($T$, °C) | $\rho\sigma$ or $\rho\sigma^+$ |
|---|---|---|---|
| $p$-$NO_2C_6H_6COO\cdot$ ($(ArCOO)_2$) | $(PhCH_2)_2O$ (italic $H_2$) | Reactants (80°–90°) | $-1.4\sigma$ |
| $PhCOO\cdot$ ($(PhCOO)_2$) | $(PhCH_2)_2O$ (italic $H_2$) | Reactants (80°–90°) | $-0.65\sigma^+$ |
| $p$-$MeOC_6H_4CH_2\cdot$ ($ArCH_2N_2CH_2Ar$) | $(PhCH_2)_2O$ (italic $H_2$) | Reactants | 0 |
| $Bu^tO\cdot$ ($Bu^t_2O_2$) | $(PhCH_2)_2O$ (italic $H_2$) | Reactants (110°) | $-0.6\sigma$ |
| $Bu^tO\cdot$ ($Bu^t_2O_2$) | PhO*H* | $CCl_4$ (122°) | $-1.19\sigma^+$ |
| $Bu^tO\cdot$ ($Bu^t_2O_2$) | PhO*H* | PhCl (122°) | $-0.74\sigma^+$ |
| $RO_2\cdot$ (RH, $O_2$) | PhO*H* | styrene (65°) | $-1.49\sigma^+$ |
| DPPH· | PhO*H* | $C_6H_6$ (30°) | $-2.77\sigma^+$ |
| DPPH· | PhO*H* | $CCl_4$ (20°) | $-4.54\sigma^+$ |
| $Bu^tO\cdot$ ($Bu^tO—C_2O_4—OBu^t$) | PhOC*H*$_3$ | $CCl_2FCClF_2$ (45°) | $-0.41\sigma^+$ |
| $Cl_3C\cdot$ ($BrCCl_3$) | PhC*H*O | $CCl_4$ (40°) | $-1.13\sigma$ |
| $Cl_3C\cdot$ ($BrCCl_3$) | PhC*H*O | $CCl_4$ (80°) | $-0.74\sigma$ |
| $ClO_2S\cdot$ ($SO_2Cl_2$) | PhC*H*O | $CCl_4$ (40°) | $-0.53\sigma$ |
| $Cl_3C\cdot$ ($BrCCl_3$) | PhCH=CHPh | reactants (105°) | $-0.7\sigma^+$ |
| $p$-$MeOC_6H_4\cdot$ | *Ph*H | $C_6H_6$ (20°) | $0.09\sigma$ |
| $Ph\cdot$ | *Ph*H | $C_6H_6$ (20°) | $0.05\sigma$ |
| $p$-$ClC_6H_4\cdot$ | *Ph*H | $C_6H_6$ (20°) | $-0.27\sigma$ |
| $p$-$NO_2C_6H_4\cdot$ | *Ph*H | $C_6H_6$ (20°) | $-0.7\sigma$ |
| $p$-$MeC_6H_4CO_2\cdot$ ($(ArCO_2)_2$, $CuCl_2$) | *Ph*H | MeCN (60°) | $-1.28\sigma^+$ |
| $PhCO_2\cdot$ ($(PhCO_2)_2$, $CuCl_2$) | *Ph*H | MeCN (60°) | $-1.61\sigma^+$ |
| $p$-$NO_2C_6H_4CO_2\cdot$ ($(ArCO_2)_2$, $CuCl_2$) | *Ph*H | MeCN (60°) | $-2.25\sigma^+$ |

* Data from ref. 13 and 15.
† The centre undergoing reaction is indicated by italics.

scant experimental evidence in its support. Among the newly available pieces of evidence may be quoted the dehydrogenation of dibenzyl ether by benzoyloxy and *p*-nitrobenzoyloxy radicals for which the $\rho$ values have been found to be $-0.65$ and $-1.4$, respectively.[5b,14] The latter radicals being more electrophilic clearly involves a greater degree of control by polar effects. Analogous results are obtained for aromatic substitution by aryl and substituted benzoyloxy radicals (Table 9.4), the electron-withdrawing nitro group obviously increasing the electron affinity of the radical. Similar trends are observed in aromatic substitution by other radicals.[15] The weakly nucleophilic methyl and *p*-methoxybenzyl, and the relatively neutral phenyl radicals, all give zero or near zero $\rho$ values in their hydrogen abstraction or aromatic substitution reactions.

Worthy of note are the remarkably large $\rho$ values, of magnitude comparable to those for ionic reactions, observed for hydrogen abstraction from phenols by the diphenylpicrylhydrazyl radical (DPPH·). This has been attributed to the effect of the nitro groups in the radical as well as its low reactivity, and to the high polarizability of the OH bond.

(b) *Reactivity of the attacking radical*

From the data shown in Tables 9.3 and 9.4 it is clear that except for closely related radicals a direct comparison of radical polarity using $\rho$ values is not possible. As has been explained previously, this is because $\rho$ values are dependent on the extent of bond reorganization in the transition state, which of course will not be the same for radicals of widely different reactivities. For a given system under attack by different radicals, the reaction involving the more reactive radical is less affected by polar effects, and vice versa. The relative $\rho$ values, at the same temperature range of 70–80 °C, for hydrogen abstraction from toluene by atomic chlorine and by atomic bromine, namely, $-0.66$ and $-1.36$, support this. Hydrogen abstraction by the more reactive chlorine atom is strongly exothermic, the transition state occurring early in the reaction coordinate (i.e., resembles the reactants) and consequently there is little bond breaking in the transition state. Abstraction by bromine, on the other hand, involves a greater extent of bond breaking in the transition state, and is thus more susceptible to polar effects, even though bromine is less electrophilic than chlorine.

A comparison of the polarity of atomic chlorine and the *t*-butoxy radical would be more appropriate here since these have similar reactivities (the latter is slightly more selective). At 40 °C we find that for hydrogen abstraction from toluene by atomic chlorine the $\rho$ value is more negative than that for the *t*-butoxy radical (from *t*-butyl peroxalate). Hence the former is probably more electrophilic.

Other conditions being the same, participation of polar effects is seen to decrease with increase in the reaction temperature, and vice versa. The results for the bromination of toluene (Table 9.3) and the more recent work by Lee[16] on the reaction of the trichloromethyl radical with benzaldehye substantiate this generalization. A greater polar effect at lower temperature is due to a corresponding decrease in reactivity of the radical.

### (c) *Reactivity of the system attacked*

The relationship between the extent of control by polar effects and the reactivity of the system under attack has been studied by Huang and Lee[17] who after examining the $\rho$ values obtained for the bromination of a series of $\alpha$ substituted toluenes and correlating these with the relative reactivities of the systems concerned drew attention to the consistent relationship which exist between these quantities. The source of atomic bromine, the hydrogen abstracting radical, was *N*-bromosuccinimide; the $\alpha$ substituted toluenes investigated included hydrocarbons, ethers, and acetals and may be represented by PhCHRR′, R, and R′ being H, Me, Ph, OMe, etc., as given in Table 9.4. The relative reactivities of the systems undergoing attack was determined by allowing pairs of the substrates, under identical conditions as employed in the Hammett studies, to compete for an insufficient quantity of atomic bromine, and analysing the reaction products or the unreacted substrates, the reactivity of each system being then related to that of toluene, taken as unity. An inspection of the values given in Table 9.5 quickly reveals the consistent relationship mentioned above, namely, that the increasing values in the reactivity column are associated with a decreasing magnitude in the values of $\rho$ in the next column. The conclusion therefore emerges that, with a given attacking radical under the same reaction conditions, control by polar effects decreases with increasing reactivity of the reacting system, and vice versa.

Since for closely related systems (such as those discussed above) the relative rate of abstraction, i.e., the relative reactivity, will correlate with the activation energy, the *A*-factors (entropy terms) of the Arrhenius equation being similar, a high reactivity value will be associated with low activation energy. This in turn means a decrease in bond breaking in the transition state, and is supported, incidentally, by the observed kinetic isotope effect $k^{H}/k^{D}$ of 4.86, 2.67, and 1.81 for the reaction of atomic bromine with the $\alpha$ hydrogen atoms in toluene, ethylbenzene, and isopropylbenzene (cumene), respectively. Thus for closely related systems under identical reaction conditions, it may be concluded, provided certain reasonable assumptions are made, that a reaction series with a lower activation energy will be associated with a smaller $\rho$ value, i.e., less subject to polar effects, and vice versa.

**Table 9.5** Correlation of Hammett Reaction Constants with Relative Reactivities (per $\alpha$ hydrogen atom) of $\alpha$ Substituted Toluenes towards Atomic Bromine in Carbon Tetrachloride at 80 °C*

| Substrates | Reactivity Value | $\rho\sigma$ or $\rho\sigma^+$ |
|---|---|---|
| $PhCH_3$ | (1) | $-1.38\sigma^+$ |
| $Ph_2CH_2$ | 18 | $-0.93\sigma$ |
| $(PhCH_2)_2$ | 22 | $-0.92\sigma$ |
| $PhCH_2CH_3$ | 23 | $-0.86\sigma^+$ |
| $PhCH_2CH{=}CH_2$ | 25 | $-0.68\sigma^+$ |
| $PhCHMe_2$ | 50 | $-0.38\sigma$ |
| $PhCH(OMe)_2$ | 53 | $-0.38\sigma^+$ |
| $PhCH_2OMe$ | 54 | $-0.35\sigma^+$ |
| $(PhCH_2)_2O$ | 79 | $-0.12\sigma^+$ |
| $Ph_2CHOMe$ | 99 | 0 |

* Data from ref. 17*c*.

(d) *Solvent effects*

Change of solvent has been known to bring about a change in $\rho$ and this might be due to a change in the polarizability of the solvent or to its complexing with the reacting radicals.[15] It would seem reasonable to expect an increase in polarizability or dielectric constant of the solvent to stabilize a charged transition state and thereby increase the value of $\rho$, particularly for a reaction following $\sigma^+$ constants. When a radical complexes with the solvent its polarity and reactivity will change and the $\rho$ value will similarly be affected. Radicals such as atomic chlorine and alkoxy radicals have indeed been known to complex with aromatic solvents. Although no systematic study of the effect of solvents on $\rho$ values has yet been made, data in Tables 9.2 and 9.3 bearing on this indicates that such effects may be important for some reactions. Consequently the more recent studies of Hammett correlations have usually been conducted with dilute solutions rather than in mixtures of the 'neat' reactants.

## 9.3 Steric effects

As has been established in ionic reactions steric effects may hinder or facilitate reactions, so it is in radical reactions. However, probably due in part to such effects being less amenable to quantitative assessment than polar effects, no systematic study has yet been carried out on their participation in radical reactions.

Steric retardation is actually a common phenomenon in radical reactions, and probably occurs even in what appears to be a simple process, the addition to terminal olefinic bonds. Although initial addition of, say, atomic bromine to the terminal carbon atom is largely due to the relative stability of the incipient radical so produced, it is likely that steric factors also play a part in retarding addition to the non-terminal carbon atom. In fact, several authors have maintained that the steric factor alone can account for such anti-Markownikov additions.[2] Several other examples of steric hindrance in addition reactions can be found in the co-polymerization of styrene with olefins,[18] in which it has been shown that monosubstituted olefins, such as acrylic acid and acrylonitrile, add more readily than the 1,2-disubstituted ones, such as crotonic acid and fumaronitrile.

Benzylic side chain reactivity towards free radicals depends on benzyl type resonance developing in the transition state. Any steric hindrance to the near planarity of the benzyl fragment will thus lead to decreased reactivity in the abstraction process (Kooyman).[19] For example, the attack of trichloromethyl radicals on hexaethylbenzene is only about half as fast as that on ethylbenzene. Steric effects also have useful consequences in synthetic chemistry. For instance, the radical oxidation of *o*-isopropyltoluene leads to isopropylbenzoic acid rather than *o*-toluic acid. Mixtures of the three isomeric diisopropylbenzenes when oxidized give a mixture of acids but leave *o*-diisopropylbenzene unattacked, and this incidentally constitutes the best method of obtaining *o*-diisopropylbenzene.

Steric inhibition of resonance has recently been reported in radicals derived from certain cyclic benzyl ethers (Huang and Lee).[20] The radicals $Ph\dot{C}HOCH_2Ph$ and $Ph\dot{C}HOMe$, generated from dibenzyl ether and benzyl methyl ether, respectively, have earlier been shown to exhibit extremes in behaviour, the former undergoing extensive fragmentation to benzaldehyde and free benzyl, while the latter dimerizes and hardly breaks down at all. Explanation for this has been found in the fact that, whereas fragmentation of the former radical results in formation of the resonance stabilized benzyl radical, that of the latter would give the highly energized methyl radical. This process being energetically not favoured, the radical exists in solution until it dimerizes with a like radical.

By analogy with the dibenzyl ether radical the radicals (9) and (10), derived from the respective cyclic ethers, would be expected also to undergo fragmentation to give the aldehydic benzyl radicals shown but, contrary to expectations, give only dimers. This has been attributed to the fact that delocalization of the unpaired electron would not be possible in the transition state involved in fragmentation, as a result of the unfavourable geometry of the molecule. In support of this postulate it has been shown that the analogous radical (11) by contrast does break down to the corresponding benzylic radical as shown, presumably because resonance stabilization is here restored by the additional benzene ring present.

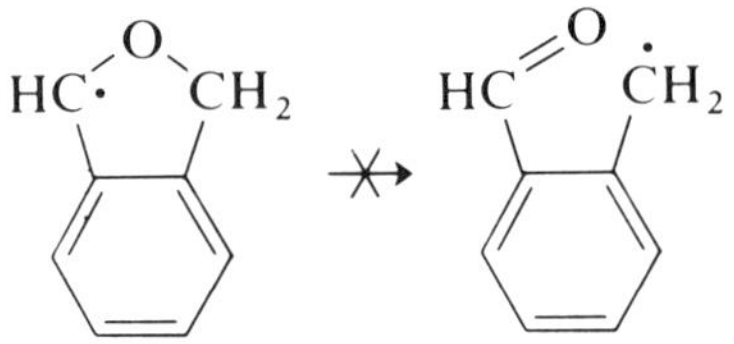

(9)

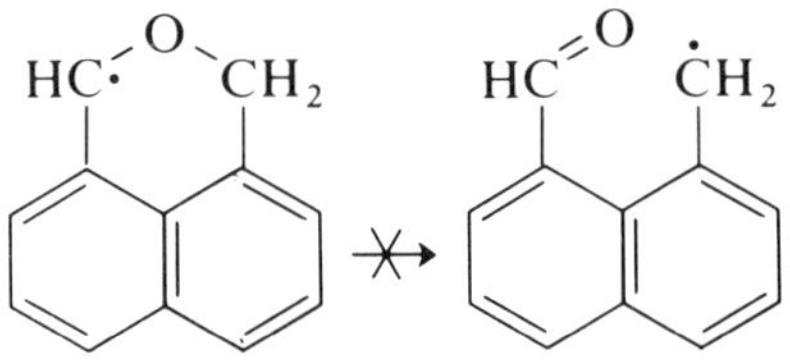

(10)

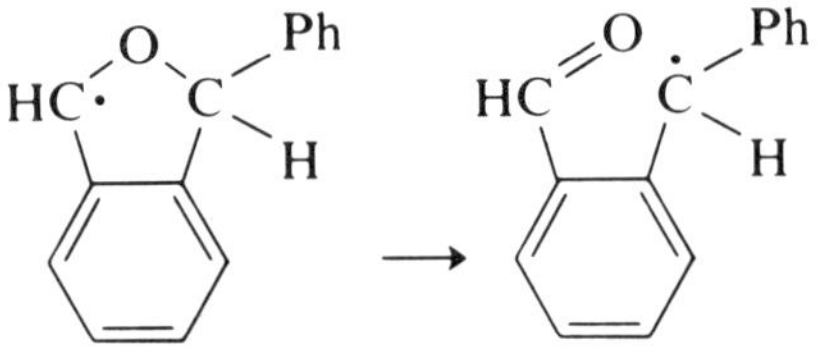

(11)

Steric effects have also been observed in the norbornyl system. Kooyman and Vegter[21] have demonstrated that the *exo* or *endo* chlorination of norbornane is a function of the size of the chlorinating agent. Molecular chlorine gives about 70% *exo* chloride, whereas the use of sulphuryl chloride, carbon tetrachloride, or phosphorus pentachloride leads to 95% of the *exo* halide. Steric effects in the 'classical' norbornyl radical (12) could arise in two ways: (a) via the *endo* approach which is more hindered by the *endo* C-6 hydrogen than is the *exo* approach by the *syn* C-7 hydrogen and (b) via an *endo* attack which requires the hydrogen at the radical centre to eclipse the C-1 hydrogen in the transition state making it energetically less favourable.

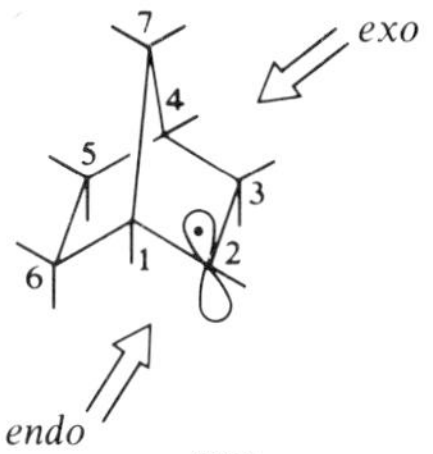

*exo*

*endo*

(12)

Another relatively recent finding worth reporting is the induced decomposition[22] of *t*-butyl peroxide by the α-alkoxybenzyl radical $\text{Ar}\dot{\text{C}}\text{HOR}$, leading to the mixed acetal ArCH(OR)(OBu$^t$) as indicated below. As seen from the results in Table 9.6 both polar effects (as exerted by Ar) and steric factors (due to the size of R) influence the extent to which induced decomposition takes place, as measured by the yield of the mixed acetal. Thus with the series $\text{Ar}\dot{\text{C}}\text{HOMe}$, only the relatively electrophilic radicals (Ar = chloro- and nitrophenyl; R = Me) are capable of entering into the reaction with the peroxide, while the neutral, unsubstituted radicals (Ar = phenyl, R = Me)

**Table 9.6** Reactions of Benzyl Radicals ArCHOR in the Presence of *t*-Butyl Peroxide*

| Ar | R | ArCHO | (ArCHOR)$_2$ | ArCH(OR)(OBu$^t$) |
|---|---|---|---|---|
| Ph | Me | 10† | 88† | 3† |
| *p*-ClC$_6$H$_4$ | Me | 10 | 60 | 36 |
| *m*-ClC$_6$H$_4$ | Me | 10 | 63 | 17 |
| *p*-NO$_2$C$_6$H$_4$ | Me | 13 | 0 | 85‡ |
| 3,4-Cl$_2$C$_6$H$_3$ | Me | 44 | trace | 39 |
| *p*-PhC$_6$H$_4$ | Me | 5 | 87 | 0 |
| *p*-MeOC$_6$H$_4$ | Me | 3 | 90 | 0 |
| *p*-Bu$^t$C$_6$H$_4$ | Me | 5 | 86 | 9 |
| *p*-ClC$_6$H$_4$ | Et | 6 | | 34 |
| *p*-ClC$_6$H$_4$ | Pr$^i$ | 12 | | 2 |
| 3,4-Cl$_2$C$_6$H$_3$ | Et | 73 | | 13 |
| 3,4-Cl$_2$C$_6$H$_3$ | Pr$^i$ | 49 | | 1 |
| 3,4-Cl$_2$C$_6$H$_3$ | Ph | 21 | | 0 |

* Data from ref. 22.

† Products expressed in % yield.

‡ 36% as the ester ArCOOMe.

and the weakly nucleophilic ones (Ar = *p*-*t*-butylphenyl and *p*-methoxyphenyl; R = Me) hardly cause any induced decomposition, but end up as the dimers. The steric consequence of the size of R is also clear. Thus in the 3,4-dichlorophenyl series changing R successively from methyl to ethyl, isopropyl and phenyl in the radical results in drastic decreases in the yield of the mixed acetals, from 39 to 13, 1, and 0% respectively.

$$ArCH_2OR \rightarrow Ar\dot{C}HOR \begin{cases} \rightarrow ArCHO + R\cdot \\ \rightarrow (ArCHOR)_2 \\ \xrightarrow{Bu^t_2O_2} ArCH(OR)(OBu^t) + Bu^tO\cdot \end{cases}$$

## 9.4 Solvent effects

Although in general solvents appear to have little effect on the course of free radical reactions when compared with ionic reactions, Russell's classical investigations[23] on the complexing of atomic chlorine with basic solvents have shown that under certain circumstances solvent effects can in fact become quite profound. Besides complexing with the attacking radical and thus modifying its properties, a solvent may affect reactions through one of its physical properties, e.g., polarizability or viscosity, or through a sheer diluent effect alter the rates at which reactions proceed.

In the photochlorination of the branched chain hydrocarbon 2,3-dimethylbutane the selectivity of atomic chlorine for tertiary over primary hydrogen atoms was found to be increased in aromatic solvents and in carbon disulphide (Table 9.7). Russell suggested that atomic chlorine forms $\pi$-complexes with these solvents and being thus stabilized becomes more selective. A few observations supporting this idea may be quoted: (a) an increase in solvent concentration increases the selectivity, (b) the effectiveness in this regard of substituted aromatics increases with their electron donating capacity, as for example in the order observed for anisole, benzene, and nitrobenzene, (c) changes in selectivity with temperature are more pronounced in aromatic solvents than in the pure hydrocarbon, and (d) the kinetic deuterium isotope effect ($k^H/k^D$) for the abstraction of tertiary hydrogen atoms from isobutane at $-15\,°C$ increases from 1.2–1.5 to 1.5–2.1 when conducted in 5.9M chlorobenzene solution.

The increased selectivity due to complexing can be explained by reference to Fig. 9.2 which shows that complexing lowers the energy content of

**Table 9.7** Solvent Effects on Photochlorination of 2,3-dimethylbutane*

| Solvent (concentration, M) | *t/p* (25 °C)† | *t/p* (55 °C)† |
|---|---|---|
| 2,3-Dimethylbutane (7.6 M) | 4.2 | 3.7 |
| Carbon tetrachloride (4.0 M) | | 3.5 |
| Carbon disulphide (2.0 M) | 15 | |
| Carbon disulphide (8.0 M) | 106 | |
| Carbon disulphide (12.0 M) | 225 | |
| Nitrobenzene (4.0 M) | | 4.9 |
| Benzene (2.0 M) | 11 | 8.0 |
| Benzene (4.0 M) | 20 | 14.6 |
| Benzene (8.0 M) | 49 | 32 |
| Chlorobenzene (4.0 M) | | 10 |
| Anisole (4.0 M) | | 18.4 |
| *p*-Xylene (4.0 M) | | 19 |
| Mesitylene (4.0 M) | | 25 |

* From ref. 23.
† Relative abstraction of tertiary to primary hydrogens, statistically corrected.

atomic chlorine and hence increases the activation energy for abstraction of hydrogen. As noted earlier complexing also reduces the polarity of the radical and hence affects the Hammett $\rho$ value but this effect is complicated by an increase of bond breaking in the transition state. Extension of the study to the *t*-butoxy radical has shown that with benzene as solvent complexing of the radical hinders its abstraction of hydrogen atoms and as a result fragmentation to acetone and free methyl radicals occurs to a larger extent.

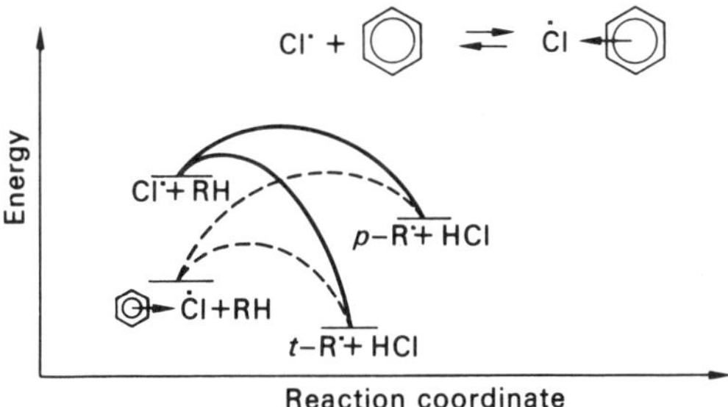


**Fig. 9.2** Hydrogen abstraction of tertiary and primary hydrogens by free and complexed chlorine atoms.

**Table 9.8** Effect of Solvents on the Behaviour of PhĊHOMe

| Solvent | Viscosity $\eta^{23} \times 10^2$ poise | Dimer: benzaldehyde Ratio |
|---|---|---|
| Pentane | 0.22 | 2.4 |
| Octane | 0.51 | 3.2 |
| Benzyl methyl ether | 1.0 | 4.2 |
| Hexadecane | 3.6 | 5.9 |
| Diphenyl ether | 3.9 | 1.3 |
| Diphenylsulphide | — | 0.12 |

Recent investigation[8] of the behaviour of the α-methoxybenzyl radical in a number of solvents shows that solvents possessing donor properties (such as diphenyl ether or diphenyl sulphide) increase the extent of fragmentation of the radical relative to dimerization, while viscous solvents (e.g., hexadecane) favour dimerization (see Table 9.8).

Much less is known of the role played by solvents in the solvation of radicals and of the transition states.[24] In ionic reactions the effects of solvents can usually be predicted with some degree of accuracy: for instance, strongly polar solvents will accelerate a reaction which is accompanied by charge formation. In radical reactions, the effect of polar solvents while not expected to be dominant is not negligible, especially in reactions of relatively polar radicals. The peroxy radical ($R—O—O\cdot \leftrightarrow R—\overset{+\cdot}{O}—\overset{-}{O}$) for example, may be expected to be somewhat solvated in the ground state but when it proceeds to the transition state of a reaction a certain degree of desolvation will have to occur. In studies of the oxidation of hydrocarbons considerable changes of kinetic parameters have in fact been observed when the medium used was varied from the non-polar hydrocarbons to the more polar solvents like acetonitrile or nitromethane. In unimolecular homolyses of peroxides and other initiators similar solvation has also been found to be of some importance.

The effect of solvents capable of forming hydrogen bonds to radicals is usually to lower the reactivity of the radical.[15b] The decreased reactivity is steric in nature since the solvating molecule will hinder the reaction of the radical with other molecules.

## References

1 R. L. Huang, *J. chem. Soc.*, 1956, 1747; 1957, 1342.
2 C. Ruchardt, *Angew. Chem., Int. ed.*, 1970, **9**, 830.
3 R. S. Davidson, *Q. Rev. chem. Soc.*, 1967, **21**, 249.

4 M. S. KHARASCH, P. C. WHITE, and F. R. MAYO, *J. org. Chem.*, 1938, **3**, 33.
5*a* R. L. HUANG and S. SI-HOE, *J. chem. Soc.*, 1957, 3988;
*b* R. L. HUANG, H. H. LEE, and S. H. ONG, *J. chem. Soc.*, 1962, 3336.
6 J. M. TEDDER, *Q. Rev. chem. Soc.*, 1960, **14**, 336.
7 M. L. POUTSMA, *Methods in Free-Radical Chemistry*, Vol. I, E. S. HUYSER, Ed., Marcel Dekker, New York, 1969, Chapter 3.
8 R. L. HUANG, T. W. LEE, and S. H. ONG, *J. chem. Soc.* (*C*), 1969, 40.
9 C. WALLING, *Free Radicals in Solution*, J. Wiley, New York, 1957, p. 132.
10*a* R. VAN HELDEN and E. C. KOOYMAN, *Recl. Trav. Chim.*, 1954, **73**, 269;
*b* C. WALLING and B. MILLER, *J. Am. chem. Soc.*, 1957, **79**, 4181.
11 G. A. RUSSELL, *J. org. Chem.*, 1958, **23**, 1407.
12 R. E. PEARSON and J. C. MARTIN, *J. Am. chem. Soc.*, 1963, **85**, 3142.
13 C. WALLING and J. A. MCGUINESS, *J. Am. chem. Soc.*, 1969, **91**, 2053.
14 R. L. HUANG, H. H. LEE and A. H. G. SIM, unpublished results, 1963.
15*a* J. A. HOWARD and K. U. INGOLD, *Can. J. Chem.*, 1963, **41**, 1744;
*b* YU. L. SPIRIN, *Russ. Chem. Rev.*, 1969, **38**, 529;
*c* M. E. KURZ and M. PELLEGRINI, *J. org. Chem.*, 1970, **35**, 990;
*d* I. B. AFANASEV, *Russ. Chem. Rev.*, 1971, **40**, 216.
16 K. H. LEE, *Tetrahedron*, 1970, **26**, 2041.
17*a* R. L. HUANG and K. H. LEE, *J. chem. Soc.* (*C*), 1964, 5963; 1966, 935;
*b* E. P. CHANG, R. L. HUANG, and K. H. LEE, *J. chem. Soc.* (*B*), 1969, 878;
*c* T. P. LOW and K. H. LEE, *J. chem. Soc.* (*B*), 1970, 535.
18 Ref. 9, p. 127.
19 E. C. KOOYMAN, *Rec. Chem. Prog.*, 1964, **25**, 3.
20 R. L. HUANG and H. H. LEE, *J. chem. Soc.*, 1964, 2500; 1966, 929.
21 E. C. KOOYMAN and G. C. VEGTER, *Tetrahedron*, 1958, **4**, 382.
22*a* R. L. HUANG, T. W. LEE, and S. H. ONG, *J. chem. Soc.* (*C*), 1969, 2522;
*b* S. H. GOH, R. L. HUANG, S. H. ONG, and I. SIEH, *J. chem. Soc.* (*C*), 1971, 2282.
23 G. A. RUSSELL, *J. Am. chem. Soc.*, 1957, **79**, 2977; 1958, **80**, 4987.
24 E. S. HUYSER, *Adv. Free-Radical Chem.*, 1965, **1**, Chapter 3, G. H. WILLIAMS, Ed., Logos, London.
25 A. A. ZAVITSAS and J. A. PINTO, *J. Am. chem. Soc.*, 1972, **94**, 7390. These authors have pointed out that Hammett studies of the hydrogen abstraction from substituted toluenes have usually neglected the effect of the substituents on the bond dissociation energies of the benzylic C—H bonds. Consequently polar effects as measured this way become overemphasized.

# 10

# Hydrogen Abstraction Reactions

The abstraction of hydrogen atoms from organic compounds is a reaction of frequent occurrence with free radicals, and one characteristic of the species. In this chapter an account will be given of hydrogen abstraction by alkoxy, alkyl, phenyl, and peroxy radicals; abstraction by atomic halogens will be discussed under halogen substitution reactions (Chapter 11). The factors which have to be taken into account in the study of these reactions, including bond energy considerations, polar and steric influences, and effects due to solvents, have been discussed at some length in Chapters 3 and 9.

## 10.1 Hydrogen abstraction by alkoxy radicals

In the presence of hydrogen donors, simple alkoxy radicals react almost invariably by abstraction of hydrogen atoms from these donors; a tertiary alkoxy radical, however, may also undergo $\beta$ scission to give an alkyl radical and a ketone (see Section 14.1). In the case of the *t*-butoxy radical competition between these two processes, one yielding *t*-butanol and the other acetone, has been made the basis of a method of comparing the donor capacities of various hydrocarbons. Since fragmentation of the *t*-butoxy radical is independent of the reactivity of the solvent as a hydrogen donor, while the rate of hydrogen abstraction by the radical is directly related to this, the more reactive the solvent is, the higher would be the proportion of *t*-butoxy radicals which end up as *t*-butanol, and hence the higher the *t*-butanol-acetone ratio, and vice versa. When measured under identical conditions for different solvents, this ratio would thus give an indication of their relative reactivities. The results obtained[1] from the decomposition of *t*-butyl peroxide in a number of hydrocarbons are given in Table 10.1, the relative rate for toluene being assigned a value of unity.

As to the selectivity of the *t*-butoxy radical this has been estimated as being intermediate between those of atomic chlorine and atomic bromine, by comparing the scales of relative reactivities of primary, secondary and tertiary C—H bonds towards these radicals. (These are 1.0:12.2:44 for *t*-butoxy radicals, and 1.0:4.3:6.0 and 1:150:4250 for atomic chlorine and

**Table 10.1** Relative Rates of Hydrogen Abstraction by *t*-Butoxy Radicals at 135 °C[1]

| Hydrogen Donor | Relative Rate |
|---|---|
| Toluene | 1.0 |
| Ethylbenzene | 3.2 |
| Isopropylbenzene | 6.4 |
| *m*-Xylene | 1.2 |
| *t*-Butylbenzene | 0.1 |
| Cyclohexane | 2.0 |
| Tetralin | 7.6 |
| Diphenylmethane | 4.2 |

bromine, respectively.) The data on selectivity are consistent with the fact that, while enthalpy changes are similar for the abstraction reaction by the *t*-butoxy radical and by atomic chlorine, the rate constant for abstraction by the former is considerably lower.[2] As to its electrophilic nature this has been demonstrated in a number of Hammett studies, including the dehydrogenation of dibenzyl ethers,[3] from all of which a substantial negative $\rho$ value has been found (Section 9.2).

**Table 10.2** Relative Rates of Allylic Attack and Addition by the Radical X· on $RCH_2CH{=}CH_2$*

| X· | Allylic attack/addition |
|---|---|
| Cl· | 0.13 |
| *t*-BuO· | 30 |
| $CH_3$· | 0.25 |
| $R'O_2$· | 1.0 |
| $CCl_3$· | 0.023 |
| R'S· | very small |

* Data taken from C. Walling, *Pure and Applied Chemistry*, 1967, **15**, No. 1, pp. 69–80, IUPAC, Butterworth, London.

Between attack on allylic hydrogen atoms (route a) and addition to olefinic double bonds (route b) the *t*-butoxy radical shows an unusual preference for the former course of reaction, in contrast to other radicals (see Table 10.2).[2]

$$\mathrm{Bu^tO\cdot + RCH_2CH{=}CH_2} \begin{cases} \xrightarrow{(a)} \mathrm{R\dot{C}HCH{=}CH_2 + Bu^tOH} \\ \xrightarrow{(b)} \mathrm{RCH_2\dot{C}H{-}CH_2OBu^t} \end{cases}$$

(R = alkyl)

In tertiary alkoxy radicals containing long chain substituents intramolecular hydrogen abstraction may become an important reaction. This is illustrated below for the radical (1), generated from 2-methylhex-2-yl hypochlorite (ROCl) in cyclohexane ($C_6H_{12}$) as solvent, which yielded no less than 81% of (2), and only 8% cyclohexyl chloride and 2% *n*-butyl chloride:

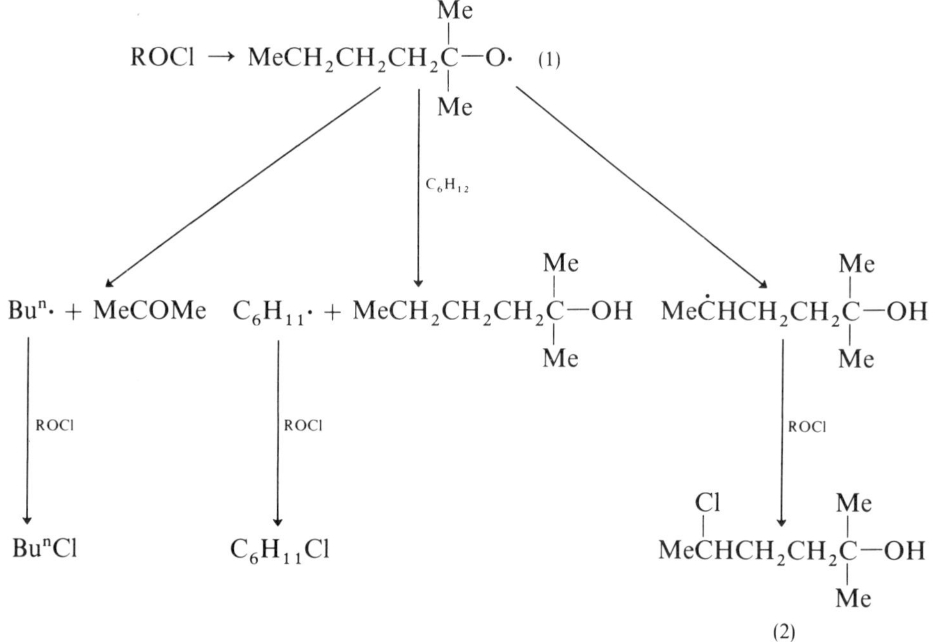


Similar cases of intramolecular halogenation in the $\delta$ position have been reported in other tertiary alkoxy radicals, and are envisaged as taking place via a quasi six-membered cyclic transition state (see also Section 7.5). Such radicals, generated in a variety of ways such as the photolysis of nitrile esters or the action of lead tetra-acetate on alcohols, have proved of wide synthetic utility.[4]

**Table 10.3** Solvent Effects on Hydrogen Abstraction from 2,3-dimethylbutane*

| Solvent | Relative Reactivities 70 °C | 0 °C |
|---|---|---|
| $PhOCH_3$ | 45 | 106 |
| PhH | 55 (40°) | 89 |
| None | 44 (40°) | 68 |
| *cis* $C_2H_2Cl_2$ | 29 | 60 |
| PhCl | 35 | 94 |
| MeCOMe | 30 | 128 |
| MeCN | 17 | 47 (25°) |

* Data from C. Walling, *Pure and Applied Chemistry*, 1967, **15**, No. 1, pp. 69–80, IUPAC, Butterworth, London.

Compared with chlorination, abstraction reactions of *t*-butoxy radicals have been found to be relatively insensitive to solvent effects (see Section 9.4), as is evident from the data in Table 10.3 showing the relative reactivities of the tertiary and primary hydrogen atoms in 2,3-dimethylbutane towards the *t*-butoxy radical in a number of solvents.[2]

## 10.2 Hydrogen abstraction by alkyl radicals

The bulk of the work on alkyl radicals has been concentrated on the methyl radical and conducted in the gas phase.[5] There is, however, convincing evidence to show that their relative reactivities are the same in solution as in the gas phase.[6] For the methyl radical abstraction of hydrogen atoms from any hydrogen donor RH which may be present in the reaction system competes with the normal termination process of dimerization. The rates for these reactions, namely,

$$\mathrm{Me}\cdot + \mathrm{RH} \xrightarrow{k_1} \mathrm{MeH} + \mathrm{R}\cdot$$

$$2\mathrm{Me}\cdot \xrightarrow{k_2} \mathrm{C_2H_6}$$

may be expressed as follows:

$$\frac{d[\mathrm{MeH}]}{dt} = k_1[\mathrm{Me}\cdot][\mathrm{RH}]$$

$$\frac{d[\mathrm{C_2H_6}]}{dt} = k_2[\mathrm{Me}\cdot]^2$$

Thus

$$\frac{d[\text{MeH}]/dt}{\{d[\text{C}_2\text{H}_6]/dt\}^{\frac{1}{2}}} = \frac{k_1[\text{RH}]}{k_2^{\frac{1}{2}}}$$

$$\frac{k_1}{k_2^{\frac{1}{2}}} = \frac{1}{[\text{RH}]}\left(\frac{\{\text{Rate of formation of methane}\}}{\{\text{Rate of formation of ethane}\}^{\frac{1}{2}}}\right)$$

The value for $k_2$ has been found to be $2 \times 10^{10}$ litre mole$^{-1}$ s$^{-1}$. Thus the rate constant, $k_1$, for the abstraction of hydrogen atoms from RH by methyl radicals is

$$k_1 = \frac{1}{[\text{RH}]}\left(\frac{\{\text{Rate of formation of methane}\}}{\{\text{Rate of formation of ethane}\}^{\frac{1}{2}}}\right) \times (2 \times 10^{10})^{\frac{1}{2}} \text{ litre mole}^{-1}\text{ s}^{-1}$$

If $k_1$ is determined over a range of temperatures, the pre-exponential factor $A$ and the activation energy $E_\alpha$ can be obtained, using the following equation

$$\ln k = \ln A - \frac{E_\alpha}{RT}$$

These quantities have been obtained for a variety of organic compounds (for the methyl and deuteromethyl radicals) and are given in Table 10.4. From these data the following observations appear of special interest:

(a) In the alkanes, the $A$ factor remains fairly constant while $E_\alpha$ decreases as the alkane becomes more branched. This is in keeping with the fact that tertiary hydrogens are abstracted more easily.

(b) Methanol and methyl ether have similar $A$ and $E_\alpha$ values, implying that hydrogen abstraction is from the methyl groups in both of these compounds.

(c) The $E_\alpha$ values for most of the aldehydes are much lower than those for methanol and methyl ether, with very small variations in the $A$ factor, showing that the aldehydes are relatively more reactive. This can be attributed to the fact that it is the labile hydrogen atom attached to the carbonyl carbon atom which suffers abstraction.

(d) The effect of substitution of deuterium for hydrogen is to increase the activation energies $E_\alpha$. (A discussion of isotope effects is given in Section 3.5).

In solution the relative reactivities of various hydrogen donors RH towards the methyl radical has been measured[7] by comparing the rate of hydrogen abstraction from RH with that of chlorine abstraction from carbon tetrachloride, added as a reference substrate. When the methyl radical is generated in a mixture of RH and $CCl_4$ the following reactions take place in competition:

$$Me\cdot + RH \xrightarrow{k_H} MeH + R\cdot$$

$$Me\cdot + CCl_4 \xrightarrow{k_{Cl}} MeCl + Cl_3C\cdot$$

**Table 10.4** Hydrogen Abstraction by Methyl and Deuteromethyl Radicals[5]

| Hydrogen Donor RH | log $A$ (mole$^{-1}$ cm$^3$ s$^{-1}$) | $E_\alpha$ (kcal mole$^{-1}$) |
|---|---|---|
| MeH* | 11.83 | 14.7 |
| EtH* | 12.21 | 11.8 |
| PrH* | 11.87 | 10.2 |
| $Bu^nH$* | 11.92 | 9.6 |
| $Me_3CH$* | 11.43 | 8.1 |
| $Me_3CH$ | 12.22 | 11.1 |
| $Me_3CD$* | 11.52 | 9.7 |
| cyclo-$C_5H_{10}$ | 12.25 | 9.1 |
| cyclo-$C_6H_{12}$ | 12.47 | 9.5 |
| $CD_3COCD_3$ | 12.07 | 11.1 |
| PhCOMe | 10.7 | 7.4 |
| $H_2CO$ | 10.25 | 6.6 |
| $D_2CO$ | 10.15 | 7.9 |
| MeCHO | 11.9 | 7.6 |
| EtCHO | 12.0 | 9.0 |
| $Pr^nCHO$ | 11.8 | 7.3 |
| $Pr^iCHO$ | 12.6 | 8.7 |
| $Bu^nCHO$ | 12.1 | 8.0 |
| $Bu^iCHO$ | 12.3 | 8.4 |
| $Bu^sCHO$ | 13.13 | 10.4 |
| $Bu^tCHO$ | 13.0 | 10.2 |
| $C\underline{H}_3NH_2$†* | 10.99 | 8.7 |
| $Me_3NH_2$† | 9.55 | 5.7 |
| $EtNH_2$ | 11.2 | 7.1 |
| $Me_2NH$ | 11.7 | 7.2 |
| $Me_3N$ | 12.6 | 8.0 |
| $C\underline{H}_3OH$† | 11.38 | 10.4 |
| MeOOMe | 12.56 | 12.0 |
| HCOOMe | 10.7 | 8.6 |

* Reactions of $CD_3$.

† When the distinction can be made the hydrogen atom abstracted is underlined.

Data from A. F. Trotman-Dickenson, *Adv. Free-Radical Chem.*, 1965, **1**, chap. 1, Logos, London.

for which the rates are:

$$\frac{d[\text{MeH}]}{dt} = k_{\text{H}}[\text{Me}\cdot][\text{RH}]$$

$$\frac{d[\text{MeCl}]}{dt} = k_{\text{Cl}}[\text{Me}\cdot][\text{CCl}_4]$$

$$\frac{d[\text{MeH}]}{d[\text{MeCl}]} = \frac{k_{\text{H}}[\text{RH}]}{k_{\text{Cl}}[\text{CCl}_4]}$$

If the amounts of methyl radicals generated are small, the concentrations of RH and $CCl_4$ remain substantially unchanged during the reaction and may be taken to be equal to their initial concentrations. Since both [MeH] and [MeCl] are initially zero, $\Delta$[MeH]/$\Delta$[MeCl] is equal to the ratio of methane to chloromethane in the products and thus,

$$\frac{k_{\text{H}}}{k_{\text{Cl}}} = \frac{[\text{CCl}_4]}{[\text{RH}]} \times \frac{[\text{yield of methane}]}{[\text{yield of chloromethane}]}$$

The results obtained by this technique, listed in Table 10.5, show that acetone and toluene react more slowly than does carbon tetrachloride while cyclohexane is an unusually reactive hydrocarbon. Octene, which has allylic hydrogens, reacts rapidly, while benzene is relatively inert.

**Table 10.5** Hydrogen Abstraction by the Methyl Radical in Solution at 100 °C[7]

| Hydrogen Donor RH | $k_H/k_{Cl}$ | Hydrogen Donor | $k_H/k_{Cl}$ |
|---|---|---|---|
| Benzene | 0.039 | 1-Octene | 3.2 |
| Methyl benzoate | 0.062 | Cyclohexane | 4.8 |
| Acetone | 0.40 | Chloroform | 11.1 |
| Toluene | 0.75 | Methyl acetate | 21 |
| Carbon tetrachloride | (1.00) | | |

Considerably less work has been done on the ethyl and higher alkyl radicals. These show behaviour similar to that of the methyl radical, except for their tendency to disproportionate. Few halogenated alkyl radicals have been studied, the more important ones being the trifluoromethyl and trichloromethyl radicals. (For abstraction reactions of the latter radical the reader is referred to Section 11.4).

## 10.3 Hydrogen abstraction by phenyl radicals

The abstraction of hydrogen atoms by phenyl radicals has been studied both in the gas phase and in solution. The latter results will be discussed in some detail below; for gas phase reactions, some typical results are summarized in Table 10.6.

**Table 10.6** Hydrogen Abstraction by Phenyl Radicals in the Gas Phase*

| Hydrogen Donor RH | $E_\alpha$ (kcal mole$^{-1}$) | log $A$ (mole$^{-1}$ cm$^3$ s$^{-1}$) |
|---|---|---|
| Molecular hydrogen | 6.5 | 10.7 |
| Methane | 11.1 | 11.9 |
| Cyclopropane | 8.5 | 11.4 |
| Isobutane | 6.7 | 11.8 |
| Acetophenone | 6.2 | 11.6 |
| Fluoroform | 5.2 | 9.9 |

* Data from A. F. Trotman-Dickenson, *Adv. Free-Radical Chem.*, 1965, **1**, chap. 1, Logos, London.

In solution the relative reactivities of a large number of hydrogen donors RH have been measured by Bridger and Russell[8] with the same competition method as described above for the methyl radical, using phenylazotriphenylmethane as the radical source. The azo compound was allowed to decompose at 60 °C in mixtures of RH and $CCl_4$ and the yield of the products determined by gas chromatography, giving

$$\frac{k_H}{k_{Cl}} = \frac{[CCl_4]}{[RH]} \times \frac{\text{(yield of benzene)}}{\text{(yield of chlorobenzene)}}$$

Some typical results so obtained[8,9] are listed in Table 10.7. Of special interest is the use of some of these data in the calculation of the relative reactivities of primary, secondary, and tertiary benzylic hydrogen atoms in aralkyl hydrocarbons which are found to be in the ratio 1.0:4.6:9.7, and in aliphatic hydrocarbons for which the ratio is 1.0:9.3:44.0. Another interesting conclusion similarly arrived at is that a primary benzylic hydrogen atom has the same order of reactivity as a secondary aliphatic hydrogen atom.

The kinetic isotope effect has been determined, by the same authors,[8] for the reaction at 60 °C of phenyl radicals with toluene and with acetone. The values of $k_H/k_D$ found (4.5 and 4.2, respectively) indicate considerable bond breaking in the transition state. A Hammett study of the toluene system, on the other hand, has given a low $\rho$ value of $-0.1$ at 60 °C, indicating relatively

**Table 10.7** Hydrogen Abstraction by the Phenyl Radical in Solution[8, 9]

| Hydrogen Donor RH | $k_H/k_{Cl}$ | Hydrogen Donor RH | $k_H/k_{Cl}$ |
|---|---|---|---|
| Dimethylpropane | 0.14 | Nitrotoluene | 0.22 |
| Hexane | 0.80 | Diphenylmethane | 1.4 |
| 2,3-Dimethylbutane | 1.19 | Triphenylmethane | 3.5 |
| Cyclohexane | 1.08 | Chloroform | 3.2 |
| Cyclohexene | 4.4 | Chloromethane | 0.07 |
| Toluene | 0.27 | Dimethyl ether | 0.28 |
| Ethylbenzene | 0.84 | Acetone | 0.17 |
| Isopropylbenzene | 0.93 | Acetic acid | 0.09 |
| *t*-Butylbenzene | 0.11 | Methyl acetate | 0.09 |

low sensitivity towards polar effects. Bridger and Russell's study further shows that the phenyl radical is similar in selectivity to the methyl radical (at 65 °), but is considerably more selective than atomic chlorine or *t*-butoxy radicals and less so than the bromine atom or the trichloromethyl radical (at 40 °).

## 10.4 Autoxidation

The term autoxidation generally refers to slow oxidative processes brought about by oxygen at moderate temperatures, as contrasted with the rapid process of combustion or inflammation which require high temperatures. After radical polymerization, autoxidation is, from the technical standpoint, probably the most important radical reaction. For instance, in the industrially important Hock and Lang Process, molecular oxygen converts cumene into dimethylbenzyl hydroperoxide in high yield, which then rearranges in acid to phenol and acetone, as follows:

$$PhCHMe_2 + O_2 \rightarrow Ph(Me)_2C{-}OOH \xrightarrow{H^{\cdot}} PhOH + MeCOMe$$

The drying of oils in paints and varnishes is another example of the autoxidation process. The action of oxygen on the drying oils produces hydroperoxides which, on decomposition in the presence of transition metal salts (see Section 8.2 and reference 10), initiate polymerization of the unsaturated systems present to produce a coating. Molecular oxygen, in fact, attacks a multiplicity of organic compounds including hydrocarbons, aldehydes, ethers, ketones, and a large number of oxygen and nitrogen containing molecules, the first stable product being a hydroperoxide. This process is the cause of deterioration of rubber and plastics, the breakdown of lubricants, and the rancidification of edible oils. The reaction is so widespread even with hydrocarbons that it has been thought to consume more of this class of compounds than all other reactions put together.[11] Effective means of minimizing or eliminating autoxidation would thus be of great economic value.

For the autoxidation of chemical compounds containing allylic hydrogen atoms (olefins, alkylbenzenes, etc.) under high oxygen pressure the pioneering work of Farmer, Bateman, Bolland, and Gee[12] reveals the following mechanism:

$$\left.\begin{array}{r} \text{Initiator} \rightarrow \text{X}\cdot \\ \text{X}\cdot + \text{RH} \rightarrow \text{XH} + \text{R}\cdot \end{array}\right\} \text{Initiation}$$

$$\left.\begin{array}{r} \text{R}\cdot + \text{O}_2 \rightarrow \text{ROO}\cdot \\ \text{ROO}\cdot + \text{RH} \xrightarrow{k_p} \text{ROOH} + \text{R}\cdot \end{array}\right\} \text{Propagation}$$

$$\left.\begin{array}{r} \text{ROO}\cdot + \text{R}\cdot \rightarrow \text{ROOR} \\ 2\text{R}\cdot \rightarrow \text{RR} \\ 2\text{ROO}\cdot \xrightarrow{2k_t} \text{ROOR} + \text{O}_2 \end{array}\right\} \text{Termination}$$

If it is assumed that the last reaction, namely, $2\text{ROO}\cdot \rightarrow \text{ROOR} + \text{O}_2$ constitutes the main termination step, then at the steady state condition the rate of oxidation would be given by the following expression:

$$\frac{-d[\text{O}_2]}{dt} = k_p \frac{[\text{RH}][R_i]^{\frac{1}{2}}}{(2k_t)^{\frac{1}{2}}}$$

that is, where $R_i$ is the rate of initiation, proportional to the propagation rate constant divided by the square root of that of the termination step.

The most fruitful approach for the determination of the relative reactivities of different substances towards peroxy radicals has been to determine $k_p$ and $k_t$ separately for each substance by one of the non-stationary state techniques, the rotating sector method or the thermocouple method.[13] Some recent values obtained for $k_p$, $k_t$, and $k_p/\sqrt{2k_t}$ are given in Table 10.8. The data show a wide variation in $k_t$ for hydrocarbons giving rise to primary, secondary, and tertiary peroxy radicals and because of this there is no correlation between the rate of oxidation ($k_p/\sqrt{2k_t}$) and the rate constant for chain propagation, $k_p$.

Polar effects have been found to be important in autoxidation. This is evidenced by the influence that *meta* and *para* substituents exert on the reaction rates. The negative $\rho$ values, correlated with $\sigma^+$, obtained for all the reactions listed in Table 10.9 mean that these reactions are all facilitated by

**Table 10.8** Oxidation of Typical Hydrocarbons Containing Allylic Hydrogen Atoms at 20 °C*

| Hydrocarbon | $k_p/\sqrt{2k_t} \times 10^5$ $M^{-\frac{1}{2}}\,s^{-\frac{1}{2}}$ | $k_p$ $M^{-1}\,s^{-1}$ | $k_t$ $M^{-1}\,s^{-1}$ |
|---|---|---|---|
| *p*-Xylene | 5 | 0.84 | $1.5 \times 10^8$ |
| Ethylbenzene | 21 | 1.3 | $2.0 \times 10^7$ |
| Allylbenzene | 50 | 10 | $2.2 \times 10^8$ |
| Cumene | 150 | 0.18 | $7.5 \times 10^3$ |
| Cyclohexene | 230 | 5.6 | $3.0 \times 10^6$ |

* Data from K. U. Ingold, *Pure and Applied Chemistry*, 1967, **15**, No. 1, pp. 49–67, IUPAC, Butterworth, London.

electron donating nuclear substituents.[16] There is clearly a much larger polar effect for abstraction from phenols than from hydrocarbons, in keeping with the fact that the peroxy radical is an electrophilic entity. In all probability, therefore, the reaction proceeds via a polar transition state as shown:

$$ROO\cdot + H{:}R \rightarrow ROO^{\ominus}{:}\dot{H}R^{\oplus}$$

The autoxidation of natural rubber (which is mainly *cis* polyisoprene) has been studied[14] with a view to elucidate the mechanism and to find ways of protecting the material against this destructive process. It has been found that the absorption of only 1% by weight of oxygen is sufficient to cause degradation of the natural rubber molecule and result in loss of strength and softening of the polymer. This is a physicochemical problem of some magnitude but in recent years researches at the Natural Rubber Producers' Research Association have brought to light the more important pathways of degradation and shown that peroxy radicals not only abstract allylic

**Table 10.9** Hydrogen Abstraction by Peroxy Radicals*

| Hydrogen Donor | Temperature (°C) | $\rho$ |
|---|---|---|
| PhMe | 90 | −0.7 |
| $PhCHMe_2$ | 60 | −0.4 |
| PhOH | 65 | −1.5 |
| $Ph_2NH$ | 65 | −0.9 |

* Data from K. U. Ingold, *Pure and Applied Chemistry*, 1967, **15**, No. 1, pp. 49–67, IUPAC, Butterworth, London.

hydrogen atoms from alkenes but can also add to the olefinic bonds. Evidence for the latter reaction is found in the isolation of the cyclic peroxide-hydroperoxide (3) formed during the autoxidation of squalene. This alternative fate of the propagating peroxy radical has been invoked to explain how, in the oxidation of rubber, a scission occurs in the main chains for roughly every 20 molecules of oxygen combined (see reaction scheme)

*Scission of rubber via cyclic peroxide*

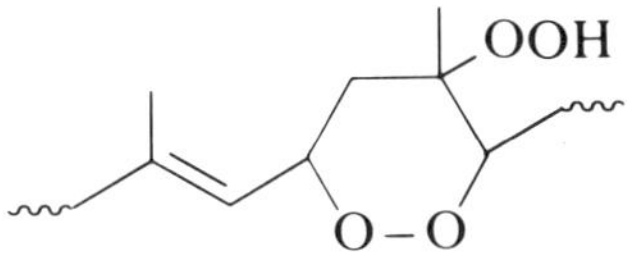

OOH

O–O

(3)

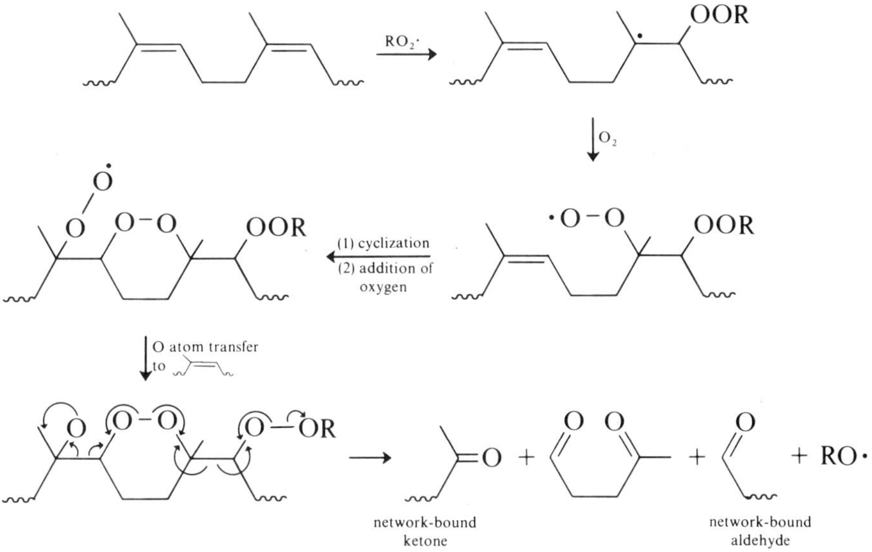

$RO_2\cdot$

OOR

$O_2$

·O–O OOR

(1) cyclization

(2) addition of oxygen

O O–O OOR

O atom transfer to

O O–O O–OR

O O O

+ RO·

network-bound ketone

network-bound aldehyde

Antioxidants used as additives to protect rubber against oxidation function in one of two ways, by harmlessly decomposing the initiating hydroperoxide or by deactivating the chain carrying peroxy radicals. The latest development in this area is the discovery of a reaction which would chemically bind an antioxidant onto the rubber network chains. Bound antioxidants have the advantage over the conventional ones in that physical losses due to solvent and aqueous leaching and to vaporization are negligible. A typical reaction leading to efficient rubber-bound antioxidants[15] is illustrated below.

*Reaction of the nitroso group*

H N O NR$_2$ → [ H O N NR$_2$ ] →

H N NR$_2$

## References


1 A. L. WILLIAMS, E. A. OBERRIGHT, and J. B. BROOKS, *J. Am. chem. Soc.*, 1956, **78**, 1190; J. H. T. BROOKS, *Trans. Faraday Soc.*, 1957, **53**, 327.

2 C. WALLING, Some Aspects of the Chemistry of Alkoxy Radicals, in *International Symposium on Free Radicals in Solution*, Butterworth, London, 1967, p. 73.

3 R. L. HUANG, H. H. LEE, and S. H. ONG, *J. chem. Soc.*, 1962, 3336.

4 K. HEUSLER and J. KALVODA, *Angew. Chem.*, 1964, **76**, 518.

5 A. F. TROTMAN-DICKENSON, *Adv. Free-Radical Chem.*, 1965, **1**, Chapter 1, G. H. WILLIAMS, Ed., Logos, London.

6 W. A. PRYOR, *Free Radicals*, McGraw-Hill, London, 1966, p. 163.

7 F. G. EDWARDS and F. R. MAYO, *J. Am. chem. Soc.*, 1950, **72**, 1265.

8 R. F. BRIDGER and G. A. RUSSELL, *J. Am. chem. Soc.*, 1963, **85**, 3754.

9 W. A. PRYOR and H. GUARD, *J. Am. chem. Soc.*, 1964, **86**, 1150.

10 W. A. WATERS, *J. Am. Oil Chem. Soc.*, 1971, **48**, 427.

11 K. U. INGOLD, Rate Constants for Some Reactions of Oxy Radicals, in *International Symposium on Free Radicals in Solution*, Butterworth, London, 1967, p. 49.

12 L. BATEMAN, *Q. Rev. chem. Soc.*, 1954, **8**, 147 and references therein.

13 G. M. BURNETT and H. W. MELVILLE, Investigation of Rates and Mechanism of Reactions, in *Technique of Organic Chemistry*, Vol. VIII, Part II, Ed. by S. L. FRIESS, E. S. LEWIS, and A. WEISSBERGER, Interscience, New York, 1963, Chapter 20.

14 L. BATEMAN, Ed., *The Chemistry and Physics of Rubber-like Substances*, Maclaren, London, 1963; P. W. ALLEN, D. BARNARD, and B. SAVILLE, *Chem. Br.*, 1970, **6**, 382.

15 M. E. CAIN, G. T. KNIGHT, P. M. LEWIS, and B. SAVILLE, Development of network bound anti-oxidants for improved ageing of natural rubber, *Proc. Natural Rubber Conference*, Kuala Lumpur, 1968.

16 See ref. 11, p. 64.

# 11

# Halogen Substitution Reactions

The great importance of halogenated compounds as starting materials for a large number of syntheses is well known. Most of these compounds can readily be obtained from the parent hydrocarbons by radical substitution reactions using elemental halogens or other halogenating agents such as sulfuryl chloride, *N*-bromosuccinimide, *t*-butyl hypohalites and polyhalogenated methanes.[1]

## 11.1 Substitution by halogens

Among the halogens, only chlorine and bromine are of suitable reactivity to be of major importance in halogen substitution reactions.[2,3] Fluorine is too reactive while iodine is almost completely inert.

The differences in reactivity of the halogens may be illustrated with the data obtained on the halogenation of methane (Table 11.1). As can be seen the fluorine reactions are highly exothermic, which explains the fact that the reactions of this halogen with organic compounds occur with great violence, and can be studied only after dilution with an inert gas. Chlorinations should be facile, the energetics being just right to promote and sustain radical chains. Chlorination of alkanes, carried out at an elevated temperature or by irradiation, has in fact been the subject of numerous studies.[4] As to bromination, since the abstraction of hydrogen atoms from alkanes by atomic bromine is endothermic, we may well expect the reaction chains to be short. Thus the quantum yield for the photobromination of cyclohexane has been found to be only two in contrast to a value as high as $10^6$ for chlorination. However, radical chain bromination of toluene is facile since the abstraction of benzylic hydrogen atoms in this case is exothermic by 3 kcal [$D(PhCH_2$—H) = 85 kcal $mole^{-1}$]. Finally, iodine does not react with alkanes, as may be expected; only at very high temperatures will these react to give dehydrogenated products.

The difference in reactivity between chlorine and bromine atoms has been discussed in Chapters 3 and 9. It may be recalled that abstraction of hydrogen atoms by atomic chlorine is relatively indiscriminate in contrast

**Table 11.1** Enthalpy Changes in the Halogenation of Methane*

| Reaction | ΔH kcal mole$^{-1}$ | | | |
|---|---|---|---|---|
| | F | Cl | Br | I |
| $X\cdot + CH_4 \rightarrow HX + CH_3$ | −32 | −1 | +15 | +33 |
| $CH_3\cdot + X_2 \rightarrow CH_3X + X\cdot$ | −70 | −23 | −21 | −18 |
| $X_2 + CH_4 \rightarrow CH_3X + HX$ | −102 | −24 | −6 | +15 |

* Values computed from bond dissociation energy data (Chapter 3, Table 3.1).

to that by atomic bromine, although the selectivity of chlorine can be increased somewhat by solvent complexing. Differences in reactivity are also shown in the halogenation of cyclohexane and toluene. Whereas towards atomic bromine toluene is 60 times more reactive than cyclohexane, towards chlorine the situation is reversed, cyclohexane being 11.2 times more reactive than toluene. Resonance stabilization by the phenyl group is apparently not important in the strongly exothermic step of hydrogen abstraction by chlorine.

Halogenation is also influenced by polar effects. Thus, the presence of trifluoromethyl, a strongly electronegative group, in an alkane exerts a directive effect on chlorination, the attack occurring at positions away from this group. Even a tertiary hydrogen atom, if flanked by two $\alpha$-$CF_3$ groups, may fail to be abstracted (see illustrations below). These and other examples in Chapter 9 demonstrate the importance of the polar effect in radical halogenation reactions.

$$CF_3CH_2CH_2CH_3 \rightarrow CF_3CH_2CH_2CH_2Cl + CF_3CH_2CHClCH_3$$

$$(CF_3)_2CHCH_3 \rightarrow (CF_3)_2CHCCl_3$$

That the stepwise chain halogenation of toluene to give the three halides is possible is also due to operation of the polar effect. In a benzyl halide the halogen atom through its radical stabilizing ability activates the C—H bond towards further halogenation but also retards this process through its polar and steric effects. There is thus a balance between these effects with the latter effects apparently dominating, so that halogenation takes place in stages as observed.

The halogenation of the cyclopropane ring is interesting. In addition to the formation of the normal substitution product, chlorocyclopropane, an addition product, 1,3-dichloropropane, is also obtained. This is apparently a displacement or $S_H2$ reaction, a unique radical process which seems to occur readily on the cyclopropane ring. It has also been found that the chlorination of nortricyclene (1) gives an addition product (2) with *exo*-2, *exo*-6 stereochemistry, indicating that the displacement of cyclopropyl carbon by atomic chlorine has occurred with inversion of configuration.

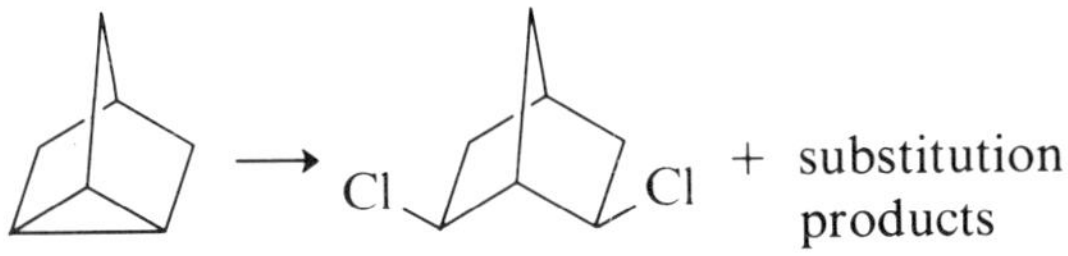

△ ⟶ △–Cl + Cl–CH₂CH₂CH₂–Cl

(1) ⟶ (2) Cl, Cl + substitution products

## 11.2 *N*-Bromosuccinimide

A mild and highly selective method of introducing bromine into an allylic position is by the use of *N*-bromosuccinimide, often abbreviated to NBS (the Wohl-Ziegler reaction).[5] The mechanism of the reaction has been a subject of controversy until recently. Bloomfield in 1944 suggested a mechanism based on the succinimidyl radical being the chain carrier, depicted as follows:

*The Bloomfield mechanism*

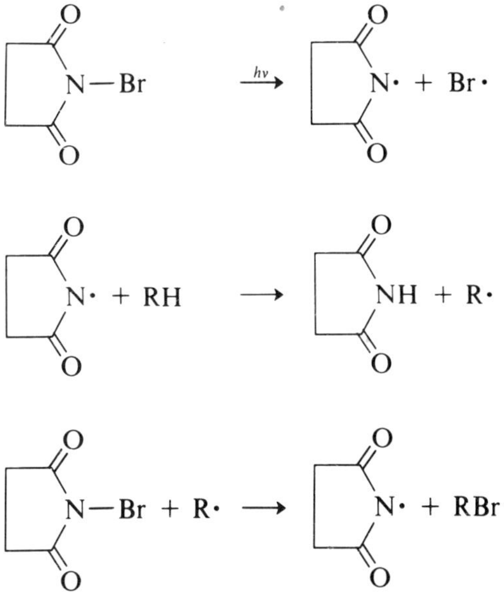

$$\text{(succinimide)N–Br} \xrightarrow{h\nu} \text{(succinimide)N}\cdot + \text{Br}\cdot$$

$$\text{(succinimide)N}\cdot + \text{RH} \longrightarrow \text{(succinimide)NH} + \text{R}\cdot$$

$$\text{(succinimide)N–Br} + \text{R}\cdot \longrightarrow \text{(succinimide)N}\cdot + \text{RBr}$$

This mechanism gained wide acceptance until a number of inconsistencies were discovered. For instance, while the reaction was found to be accelerated by radical initiators or ultraviolet light, bromination was sometimes brought about by the presence of traces of water, hydrogen bromide, molecular bromine and amines which in fact serve as bromine generators from the reagent. Goldfinger[6] in 1953 observed that *N*-bromosuccinimide behaved like *N*-chlorosuccinimide which was known to provide a source of molecular chlorine in low concentrations, and is thus in fact a positive halogen compound, i.e., one in which the molecule is strongly polarized. Reaction with traces of hydrogen bromide would readily provide molecular bromine, which is the actual brominating agent:

*The Goldfinger mechanism*

$$\text{(succinimide ring)}\overset{\delta-}{N}-\overset{\delta+}{Br} + HBr \longrightarrow \text{(succinimide ring)}NH + Br_2$$

$$Br_2 \xrightarrow{h\nu} 2Br\cdot$$

$$Br\cdot + H-R \rightarrow HBr + R\cdot$$

$$Br_2 + R\cdot \rightarrow RBr + Br\cdot$$

That selective allylic hydrogen abstraction occurs in preference to addition to the double bond is due to the very low concentration of bromine. Under such conditions the addition process, which is reversible, does not compete effectively with hydrogen abstraction which is not reversible, as shown for the allylic system (1).

$$-CH_2-\dot{C}H-CH_2Br \overset{Br\cdot}{\leftrightarrows} -CH_2-CH{=}CH_2 \xrightarrow{Br\cdot} -\dot{C}H-CH{=}CH_2$$

(1)

$$-CH_2-\dot{C}H-CH_2Br \xrightarrow{Br_2} -CH_2-CHBr-CH_2Br$$

$$-\dot{C}H-CH{=}CH_2 \xrightarrow{Br_2} -CH(Br)-\dot{C}H{=}CH_2$$

Several pieces of evidence supporting the Goldfinger mechanism have been obtained. Firstly it has been demonstrated that bromination with bromine in low concentration is similar to NBS bromination.[7] Thus allylic bromination can be achieved by the slow introduction of bromine in a stream of nitrogen into a refluxing solution of cyclohexane or methyl crotonate in carbon tetrachloride irradiated with a mercury lamp. Under such conditions it has also been found that *cis* 3-hexene isomerizes to *trans* 3-hexene, demonstrating the reversibility of the initial step in the unsuccessful competing addition reaction.

The most compelling evidence has been obtained from studies of the competitive bromination of substituted toluenes and related compounds. The benzylic hydrogens in these systems are found to have similar reactivities towards bromine and NBS; the deuterium isotope effects[3] in the bromination of toluene and deuterotoluene with both reagents are again essentially the same (see Section 3.5). Furthermore, Pearson, and Martin[8] have made Hammett studies on the bromination of substituted toluenes using molecular bromine, NBS, *N*-bromotetrafluorosuccinimide, and *N*-bromotetramethylsuccinimide (Section 9.2), and obtained nearly identical $\rho$ values, namely $-1.36$ at 80 °C (correlation with $\sigma^+$) for all these reagents. The inevitable conclusion thus emerges that, at least for benzylic bromination, hydrogen abstraction is performed by atomic bromine.

Although most brominations using NBS probably involve the agency of atomic bromine a few instances in which the succinimidyl radical participates in the reaction are known. For example the fragmentation of the succinimidyl radical to give $\beta$-bromopropionyl isocyanate takes place when NBS is allowed to react, in the presence of a radical initiator, with an olefin which has relatively unreactive allylic hydrogen atoms, as in allyl bromide. Under these circumstances bromine atoms are used up in addition reactions without producing hydrogen bromide which is required for the regeneration of molecular bromine, so that the fragmentation of the succinimydyl radical become an appreciable competing reaction. Finally, the possibility of ionic side reactions in NBS brominations must also be considered. The ionic addition of NBS to double bonds is not unknown. In another instance, compound (2), which at first sight appears to be a product of radical coupling

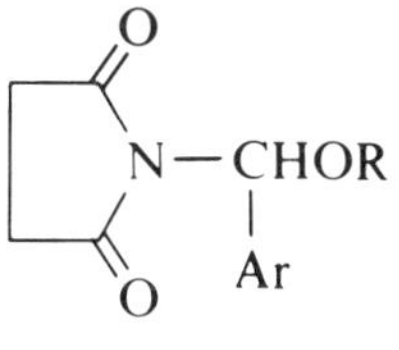

$$\text{(succinimidyl) N}\cdot \rightarrow \cdot CH_2CH_2CONCO \xrightarrow{NBS} BrCH_2CH_2CONCO$$

$$\text{(succinimido) N—CHOR(Ar)}$$

(2)

(a) Ar = Ph, R = $PhCH_2$
(b) Ar = $p$-$NO_2C_6H_4$—; R = OMe
(c) Ar = Ph, R = $p$-$BrC_6H_4$

in the reaction of NBS and several benzylic ethers, is actually formed via ionic pathways from NBS and α-bromoether intermediates.[9] However, the succinimidyl radical is known to be involved in reactions of *N*-iodo-succinimide.

## 11.3 Hypohalites

A radical chlorinating agent of great utility is *t*-butyl hypochlorite (Walling).[10] The main chain carrier from this reagent, the *t*-butoxy radical, is a rather unusual radical in that it abstracts hydrogens in preference to adding to double or triple bonds of olefins or acetylenes, thus rendering stereospecific allylic chlorinations possible, with better selectivity than

$$Bu^tO\cdot + (CH_3)(H)C{=}C(R)(H) \longrightarrow (\dot{C}H_2)(H)C{-}\dot{C}(R)(H) \xrightarrow{Bu^tOCl} (ClCH_2)(H)C{=}C(R)(H) + Bu^tO\cdot$$

$$\searrow CH_2{=}CHCHClR$$

employing molecular chlorine (see Table 11.2). In mechanistic studies it has to be noted that chlorine atom chains, in addition to the *t*-butoxy radical chains, are also operative. Thus Hammett $\rho$ values obtained for the *t*-butoxy radical from reactions using *t*-butyl hypochlorite are not reliable (Section 9.2).

**Table 11.2** Relative Selectivities of some Chlorinating Agents*

| Reagent | Chain Carrier | Selectivity† (°C) |
|---|---|---|
| $Cl_2$ | $Cl\cdot$ | *t*:*s*:*p* 4:2.5:1 (25°) |
| $Cl_2$ in $C_6H_6$ | $C_6H_6 \longrightarrow Cl\cdot$ | *t*:*s*:*p* 50–100:10:1 |
| $Bu^tOCl$ | $Bu^tO\cdot$ and $Cl\cdot$ | *t*:*s*:*p* 44:8:1 (40°) |
| $SO_2Cl_2$ | $ClSO_2\cdot$ and $Cl\cdot$ | better than $Cl_2$ |
| $SO_2Cl_2$ in $C_6H_6$ | $C_6H_6 \longrightarrow Cl\cdot$ | similar to $Cl_2$ in $C_6H_6$ |
| $CCl_3SO_2Cl$ | $CCl_3SO_2\cdot$ | *s*:*p* 25–30:1 |
| $CCl_3SCl$ | $CCl_3S\cdot$ | *t*:*s*:*p* 100:30:1 |
| *N*-chlorosuccinimide | $Cl\cdot$ | similar to $Cl_2$ |

* Data from M. L. Poutsma, *Methods in Free Radical Chemistry,* 1969, **1**, chap. 3, E. S. Huyser, Ed., Marcel Dekker, New York.

† Selectivity of C—H abstractions per hydrogen: *t*, *s*, and *p* refer to tertiary, secondary, and primary hydrogen atoms.

Atomic chlorine chains can be eliminated by the addition of small amounts of an olefin, e.g., trichloroethylene, which acts as a trap for chlorine atoms.[11]

*t*-Butyl hypobromite, because of its instability, is less important as a halogenating agent than the hypochlorite. Other hypohalites are also unstable. Acyl hypobromites, for example, are intermediates in the Hundsdiecker reaction which is useful in the preparation of alkyl bromides from carboxylic acids, as illustrated below:

$$RCOOAg + Br_2 \rightarrow RCOOBr + AgBr$$

$$RCOOBr \rightarrow RCOO\cdot + Br\cdot$$

$$RCOO\cdot \rightarrow R\cdot + CO_2$$

$$R\cdot + RCOOBr \rightarrow RBr + RCOO\cdot$$

## 11.4 Bromotrichloromethane

Of the polyhaloalkanes useful as halogenating agents bromotrichloromethane is the most important. The reagent is also employed extensively for addition reactions to unsaturated compounds (Chapter 12). It is particularly effective for benzylic brominations, the radical carrier, $Cl_3C\cdot$, being highly selective (Section 3.4). The mechanism of halogenation may be written as

$$BrCCl_3 \xrightarrow{h\nu} Br\cdot + \cdot CCl_3$$

$$Cl_3C\cdot + RH \rightarrow Cl_3CH + R\cdot$$

$$R\cdot + BrCCl_3 \rightarrow RBr + \cdot CCl_3$$

In instances where the substrate is sensitive to acids (HBr) or the electrophilic nature of bromine, bromotrichloromethane is preferred over NBS or molecular bromine, as in the benzylic bromination of *p*-methoxytoluene.[12]

Carbon tetrabromide can also serve as a brominating agent. Carbon tetrachloride with benzoyl peroxide has been used in chlorination but chain lengths are short since abstraction of chlorine atoms by carbon radicals is less facile than for bromine atoms.

## 11.5 Other halogenating agents

Several other chlorinating agents[2] are known and a few of these are given in Table 11.2. Sulfuryl chloride, which is more convenient to use than gaseous chlorine and is a particularly effective reagent, was first investigated by Kharasch and Brown in 1939 and probably involves chlorosulphonyl and chlorine radicals as chain carriers, as shown. The chlorine chain may be eliminated by adding trichloroethylene[13] as in the *t*-butyl hypochlorite reaction. In aromatic solvents the chain carrier becomes mainly complexed chlorine atoms.

$$SO_2Cl_2 \xrightarrow{h\nu} ClSO_2\cdot + Cl\cdot$$

$$Cl\cdot + RH \rightarrow R\cdot + HCl$$

$$ClSO_2\cdot + RH \rightarrow R\cdot + HCl + SO_2$$

$$R\cdot + SO_2Cl_2 \rightarrow RCl + \cdot SO_2Cl$$

Trichloromethanesulphonyl chloride, $CCl_3SO_2Cl$, is similar in action to sulphuryl chloride. Although the chain carrying radical here, $CCl_3SO_2\cdot$, is rather selective the use of the reagent is limited because radical chains are relatively short and thus large amounts of initiator are often required.

## References


1 G. SOSNOVSKY, *Free Radical Reactions in Preparative Organic Chemistry*, Macmillan, New York, 1964, Chapter 8.
2 M. L. POUTSMA, Free-Radical Chlorination of Organic Molecules, in *Methods in Free Radical Chemistry*, 1969, **1**, Chapter 3, E. S. HUYSER, Ed., Marcel Dekker, New York.
3 W. A. THALER, Free-Radical Brominations, in *Methods in Free Radical Chemistry*, 1969, **2**, Chapter 2, E. S. HUYSER, Ed., Marcel Dekker, New York.
4 C. WALLING, *Free Radicals in Solution*, Wiley, New York, 1957, p. 352.
5 L. HORNER and E. H. WINKELMANN, *Angew. Chem.*, 1959, **71**, 349.
6 J. ADAM, P. A. GOSSELAIN, and P. GOLDFINGER, *Nature*, Lond., 1953, **171**, 704.
7 B. P. MCGRATH and J. M. TEDDER, *Proc. chem. Soc.*, 1961, 80.
8*a* R. E. PEARSON and J. C. MARTIN, *J. Am. chem. Soc.*, 1963, **85**, 354;
*b* R. E. PEARSON and J. C. MARTIN, *J. Am. chem. Soc.*, 1963, **85**, 3142.
9 R. L. HUANG and K. H. LEE, *J. chem. Soc.*, 1964, 5957.
10 C. WALLING and B. B. JACKNOW, *J. Am. chem. Soc.*, 1960, **82**, 6108, 6113.
11 C. WALLING and J. A. MCGUINNESS, *J. Am. chem. Soc.*, 1969, **91**, 2053.
12 E. S. HUYSER, *J. Am. chem. Soc.*, 1960, **82**, 391.
13 K. H. LEE and T. O. TEO, *Chem. Commun.*, 1970, 860.

# 12
# Radical Addition Reactions

Partly for historical reasons, addition reactions are the most intensively studied and best understood among radical reactions. Besides being of much theoretical interest, they have assumed great practical importance by virtue of their diverse applications in organic synthesis.[1]

## 12.1 Addition of hydrogen bromide

The first radical addition reaction to receive intensive study was the 'abnormal' addition of hydrogen bromide to unsymmetrical olefins, which proceeds contrary to the Markownikov rule (see Chapter 1). The classical work of Kharasch and his co-workers[2] led to the proposal of a mechanism in 1937, still accepted today, in which atomic bromine is envisaged as the reacting entity resulting from the presence in the reaction system of oxygen, peroxide, or other substances which give rise to radicals. For allyl bromide the mechanism is represented as follows:

$$HBr + R\cdot \rightarrow RH + Br\cdot \quad (R\cdot \text{ being the initiating radical})$$

$$Br\cdot + BrCH_2{-}CH{=}CH_2 \rightarrow BrCH_2{-}\dot{C}HCH_2Br \quad (1)$$

$$(1) + HBr \rightarrow BrCH_2{-}CH_2{-}CH_2Br + Br\cdot \quad \text{etc.}$$

$$BrCH_2{-}CHBr{-}CH_2 \quad (2)$$

This radical addition, when conditions are conducive to its occurrence, is much more rapid than the ionic addition (for which the Markownikov rule holds), and thus the formation of the 1:3-dibromide takes precedence over that of the 1:2-isomer. The formation of (1), rather than (2), as the reaction intermediate finds explanation on the grounds that (1), whose radical centre is stabilized through hyperconjugation by four $\alpha$ hydrogen atoms, is more stable than (2), which is so stabilized by two $\alpha$ hydrogen atoms only.

## 12.2 Mechanism of radical addition reactions

The above mechanism has become the prototype of a large number of addition reactions to olefins. In general the reaction may be represented as the addition of XY to the olefin $R'—CH{=}CH_2$ to give $R'—CHX—CH_2Y$, mechanistically depicted as follows:

*Initiation*

$$\text{Initiator} \rightarrow R\cdot$$

$$R\cdot + XY \rightarrow RX + Y\cdot \qquad \text{(Radical initiation)}$$

or

$$XY \rightarrow X\cdot + Y\cdot \qquad \text{(Photo-initiation)}$$

*Propagation*

$$Y\cdot + R'—CH{=}CH_2 \rightarrow R'\dot{C}H—CH_2—Y\ (3) \qquad \text{(Addition)}$$

$$R'—\dot{C}H—CH_2Y + XY \rightarrow R'—CHX—CH_2Y\ (4) + Y\cdot \qquad \text{(Chain transfer)}$$

*Termination*

$$Y\cdot + Y\cdot \rightarrow Y—Y$$

$$(3) + (3) \rightarrow \text{Dimer}$$

$$(3) + Y\cdot \rightarrow R'—CHY—CH_2Y$$

*Telomerization*

$$(3) + R'CH{=}CH_2 \rightarrow R'—\dot{C}H—CH_2—CHR'—CH_2Y \text{ etc. } (5)$$

The overall reaction rate and the kinetic chain length, which essentially determine the yield of the 1:1-adduct (4), thus depend on each of the three processes of initiation, propagation and termination. These are discussed in greater detail in the following paragraphs.

### (A) *Chain initiation*

Both radical initiation and photo-initiation can be used for starting the chain sequence that leads to 1:1-adducts. In the former, benzoyl and *t*-butyl peroxides are the most frequently employed reagents, while ultraviolet rays provide a convenient energy source for the latter.

As indicated above, initiation occurs either when the radical R· from the initiator abstracts X from the addendum XY or when XY undergoes homolysis, to generate the radical Y· which then effects the first step of the sequence. There is another way in which Y· could be produced. Thus R· could add to the olefin directly and the new radical so produced then removes X from XY, as indicated:

$$R\cdot + R'—CH{=}CH_2 \rightarrow R'—\dot{C}H—CH_2R$$

$$R'—CH—CH_2R + XY \rightarrow R'—CHX—CH_2R + Y\cdot$$

Data about the initiation step are sparse. This is because in most radical chain reactions the average kinetic chain lengths are very large and consequently the incorporation into the adduct of groupings from the initiators is very small. Evidence for the abstraction reaction being mainly responsible for the initiation step is found in the *t*-butyl peroxide catalysed addition of bromotrichloromethane to stilbene, in which 64% of the peroxide appears as *t*-butyl hypobromite:

$$Bu^t_2O_2 \longrightarrow Bu^tO\cdot \xrightarrow{BrCCl_3} Bu^tOBr + \cdot CCl_3$$

However, in the benzoyl peroxide initiated addition of carbon tetrachloride to cyclohexene, 55% of the radicals from the peroxide end up in 2-chlorocyclohexylbenzoate, 10% in chlorobenzene, and 12% as benzoic acid, indicating that in this case chain initiation by addition predominates:

$$(PhCO)_2O_2 \rightarrow 2PhCO_2\cdot \rightarrow 2Ph\cdot + 2CO_2$$

$$PhCO_2\cdot + \text{cyclohexene} \longrightarrow \text{2-}(PhCO_2)\text{cyclohexyl radical} \xrightarrow{CCl_4} \text{1-}PhCO_2\text{-2-Cl-cyclohexane} + \cdot CCl_3$$

(Initiation by addition)

$$Ph\cdot + CCl_4 \rightarrow PhCl + \cdot CCl_3$$

(Initiation by abstraction)

(B) *Chain propagation*

Chain propagation consists of two steps, one of addition followed by another of chain transfer, to give the 1:1-adduct (4). Complications would arise if the intermediate radical (3), instead of undergoing transfer to carry on the chain, adds to the olefin to give a new radical (5) which by a repetition of this process leads to a chain polymer or 'telomer' $R'CHX—CH_2—(CHR'—CH_2)_n—CHR'—CH_2Y$. This side reaction, known as telomerization, would not assume serious proportions if both of the propagation steps are rapid compared with all chain termination steps. On the other hand, if either of these steps is significantly endothermic, addition by way of a free radical chain mechanism would be difficult or impossible, regardless of the efficiency of the initiators. If both propagation steps are exothermic, chain addition is possible provided the energies of activation are low. Hence it is important to consider the enthalpy changes for the two steps. An instance of such consideration, for the addition of hydrogen bromide to ethylene, is given below.

(a) $Br\cdot + CH_2 = CH_2 \rightarrow BrCH_2—CH_2\cdot$
$\pi$ bond strength = 58 kcal mole$^{-1}$; $D$(C—Br) = 67 kcal mole$^{-1}$
Hence $\Delta H$ = (58 − 67) = or −9 kcal mole$^{-1}$
(b) $BrCH_2CH_2\cdot + HBr \rightarrow BrCH_2CH_3 + Br\cdot$
$D$(H—Br) = 88 kcal mole$^{-1}$; $D(CH_3CH_2—H)$ = 98 kcal mole$^{-1}$
Hence $\Delta H$ = (88 − 98) or −10 kcal mole$^{-1}$.

As both propagation steps are exothermic, the reaction is facile.

Walling[3] has derived approximate energies of the propagation steps in a number of radical addition reactions and from these values, listed in Table 12.1, it can be predicted that radical chain processes involving the addition of molecular hydrogen, water or hydrogen fluoride are unlikely, because of the difficulty of breaking, respectively, the H—H, H—OH, and H—F bonds. The addition of iodine or hydrogen iodide is not possible for a different reason, *viz.*, the weakness of the C—I bond. In the other cases

**Table 12.1** Approximate Energies (kcal mole$^{-1}$) of Propagation Steps in Radical Additions to Ethylene*

| X—Y | $Y\cdot + CH_2{=}CH_2 \rightarrow \cdot CH_2—CH_2Y$ | $\cdot CH_2—CH_2Y + XY \rightarrow CH_2X—CH_2Y + Y\cdot$ |
|---|---|---|
| H—H | −40 | 6 |
| H—OH | −32 | 22 |
| H—F | — | 37 |
| H—SH | −16 | −8 |
| H—Br | −9 | −10 |
| H—I | 7 | −27 |
| I—I | 7 | −13 |
| $H—CH_2Ph$ | −1.5 | −21 |
| $H—COCH_3$ | −16 | −10 |
| $H—CCl_3$ | −14 | −8 |
| $Cl—CCl_3$ | −14 | −8 |

* Data taken from C. Walling, *Free Radicals in Solution*, John Wiley and Sons Inc., New York, 1957, p. 241.

where the propagation steps have been computed as exothermic, chain additions have been realized experimentally.

### (C) *Chain termination*

The way by which a chain is terminated may vary with the type of compound involved. In general chain termination occurs by radical destroying processes such as dimerization or disproportionation. It has been shown that in the addition of bromotrichloromethane to cyclohexene, chain termination is due exclusively to dimerization: at a high concentration of halide (halide-olefin ratio of 10:1), hexachloroethane has been isolated while at a low concentration (halide-olefin ratio of 1:10), 2,2′-di(trichloromethyl)-bicyclohexane is the dimer produced. Another mode of termination is the formation of resonance stabilized allyl radicals followed by dimerization of these radicals. The high yields of chloroform and dicyclohexenyl in the reaction between carbon tetrachloride and cyclohexene show the importance of this type of termination. Termination of radical chains in addition reactions

$$\cdot CCl_3 + \quad \longrightarrow \quad CHCl_3 +$$

$$2 \quad \longrightarrow$$

can also be brought about by interaction with metallic halides. It has been shown by Kochi[4] that the polymerization of styrene can be inhibited by cupric or ferric chlorides with the formation of chloro compounds as follows:

$$R\cdot + Ph{-}CH{=}CH_2 \rightarrow Ph{-}\dot{C}H{-}CH_2R$$

$$Ph\dot{C}H{-}CH_2R + FeCl_3 \rightarrow Ph{-}CHCl{-}CH_2R + FeCl_2$$

### (D) *Reversibility of the addition step*

As to whether the addition step is reversible or otherwise may be inferred from *cis–trans* isomerization which the olefins may undergo during the reaction. Isomerization in all likelihood arises from free rotation within the intermediate radical (as shown in the reaction scheme), provided this occurs faster than its reaction with the addendum XY. While this proviso is not always fulfilled, where an olefin is found to be converted into the thermodynamically more stable form (usually *cis* to *trans*) the isomerization

$$\mathrm{Y}\cdot + \underset{cis}{\mathrm{R'(H)C{=}C(H)R'}} \rightleftarrows \mathrm{[Y;\ R',\ H;\ R',\ H]} \xrightarrow{\mathrm{XY}} \mathrm{[Y;\ R',\ H;\ R',\ H;\ X]}$$

$$\Big\updownarrow \text{fast}$$

$$\mathrm{Y}\cdot + \underset{trans}{\mathrm{R'(H)C{=}C(H)R'}} \rightleftarrows \mathrm{[Y;\ H,\ R';\ R',\ H]} \xrightarrow{\mathrm{XY}} \mathrm{[Y;\ H,\ R';\ R',\ H;\ X]}$$

may be taken as indicating reversibility in the addition step. Such reversibility has been demonstrated in a few instances. Thus atomic bromine has been reported to cause isomerization of *cis* to *trans* but-2-ene (Tedder).[5] Again, in the addition of ethyl cyanoacetate to *cis* cinnamonitrile, the nitrile which was recovered was found to contain no less than 88% of the *trans* isomer (Huang, Ong, and Ong).[6] This is understandable as the propagation step in this reaction is known to be relatively slow. Similarly, *cis–trans* isomerization of natural rubber (mainly *cis* polyisoprene) can be induced by either thiol acid or sulphur dioxide.

## 12.3 Addition of substances other than hydrogen bromide

Free radical chain addition to carbon-carbon multiple bonds to form carbon-carbon and carbon-heteroatom bonds is an important synthetic method of wide scope. Compounds which undergo radical addition to olefins include polyhalogenomethanes, aldehydes, ketones, esters, alcohols, hydrocarbons, mercaptans, and other sulphur compounds, certain phosphorus and silicon compounds, and a few derivatives of the less common elements. It is of interest to note, however, that addition to a carbon-heteroatom multiple bond is seldom observed. One of the few known cases of addition to a carbonyl group is that involving biacetyl (Bentrude and Darnall),[7] in the synthesis of acetylcyclohexane outlined below:

$$\mathrm{C_6H_{12}} \xrightarrow{\mathrm{R}\cdot} \mathrm{C_6H_{11}}\cdot \xrightarrow{\mathrm{(MeCO)_2}} \mathrm{C_6H_{11}}\overset{\overset{\displaystyle \mathrm{O}\cdot}{|}}{\mathrm{C}}\mathrm{MeCOMe} \xrightarrow{-\mathrm{MeCO}\cdot} \mathrm{C_6H_{11}COMe}$$

Another instance, observed earlier, is the addition of the benzoyl radical to benzaldehyde (see below).

(A) *Addition of halogenated methanes*

In 1947 Kharasch, Jensen, and Urry reported[8] the addition of a number of halomethanes, namely, carbon tetrachloride, carbon tetrabromide, bromoform, and chloroform, to olefins such as 1-octene, styrene, and ethylene. This work has led to extensive study on the addition of halomethanes to olefins in general, and has proved to be of great theoretical interest and wide synthetic application.[1]

Among halides found to undergo addition are $CBr_4$, $CCl_4$, $BrCCl_3$, $BrCHCl_2$, $CHBr_3$, $CHCl_3$, $CHI_3$, and $ICF_3$. Halogenated methanes containing less than three halogen atoms do not add readily, except where other unsaturated groups are present, as in $BrCH_2COOEt$ and $ClCH_2COCH_3$.

The reaction is carried out conveniently by either (a) illuminating a mixture of the olefin and the halomethane with ultraviolet rays, or (b) adding a radical initiator to the mixture, the most commonly used initiator being benzoyl peroxide (Kharasch and co-workers used acetyl peroxide in their earlier experiments) usually at a temperature of between 60–80 °C. The mechanism operative in the addition of bromotrichloromethane, which is typical of this group of reactions, is given below:

$$\left.\begin{array}{l} BrCCl_3 \rightarrow Br\cdot + \cdot CCl_3 \\ \text{or} \\ BrCCl_3 + R\cdot \rightarrow R{-}Br + \cdot CCl_3 \end{array}\right\} \quad \text{initiation}$$

$$R'{-}CH{=}CH_2 + \cdot CCl_3 \rightarrow R'{-}\dot{C}H{-}CH_2CCl_3 \ (4) \quad \text{addition}$$

$$(4) + BrCCl_3 \rightarrow R'{-}CHBr{-}CH_2CCl_3 + \cdot CCl_3 \quad \text{transfer}$$

That $BrCCl_3$ adds as Br and $CCl_3$ follows from the initiation step, in which the weakest bond, the C—Br bond, is severed, releasing the $\cdot CCl_3$ radical which adds to the olefinic double bond. In $CHBr_3$, the severance of the C—Br, rather than the C—H, bond would be expected, and this reagent thus adds as Br and $CHBr_2$ as has been observed. However, $CHCl_2$ adds as H and $CCl_3$, despite the fact that the C—H bond is a stronger bond than C—Cl. This can be attributed to the greater resonance energy of the $Cl_3C$ radical compared to the $Cl_2CH$ radical, the initiation step being $CHCl_3 \rightarrow H\cdot + Cl_3C\cdot$ rather than $CHCl_3 \rightarrow Cl\cdot + \cdot CHCl_2$. Actually, the balance must be rather close, since in the addition of $CDCl_3$, in which the rate of C—D cleavage is reduced about 11–12 fold due to the isotopic effect, cleavage of the C—Cl bond occurs to a significant extent, resulting in $CDCl_3$ adding in part as Cl and $CDCl_2$, and in part as D and $CCl_3$ (as in $CHCl_3$). In $ICF_3$ the bond severed is, as expected, the C—I bond, and this reagent adds as I and $CF_3$.

As indicated earlier for radical addition reactions in general, the yield of the 1:1-adduct depends on a number of factors. Some experimental findings are quoted below to illustrate this.

(a) In addition to ethylene, $CCl_4$ and $CHCl_3$ produce appreciable quantities of telomers at the expense of the 1:1-adduct, whereas $BrCCl_3$ gives hardly any telomer. In the last named addendum the chain transfer step involves the breaking of the weaker C—Br bond and is therefore relatively efficient.

(b) In addition to styrene, $CCl_4$ gives only telomers even with a 50 molar excess of the addendum, whereas $CBr_4$ affords a 96% yield of the 1:1-adduct with only a 6 molar excess. An equimolecular mixture of styrene and carbon tetrachloride, when acted upon by radicals generated from benzoyl peroxide, in fact gives rise to a telomeric product of the structure $Cl_3C(CH_2—CHPh)_nCl$ of which the average molecular weight and percentage of chlorine indicate that n = 70. The intermediate radical here, $Ph\dot{C}HCH_2CCl_3$, is resonance stabilized, and with the addendum being carbon tetrachloride with which the transfer step involves breaking of the C—Cl bond, the chain completely breaks down, giving no 1:1-adduct at all. Transfer constants would thus be expected to vary with the C—X bond to be broken and to increase with changes in X in the order H, Cl, Br, and I, as has been found to be the case. Bromotrichloromethane, being a relatively efficient transfer agent, has thus proved to be one of the most versatile reagents for study of various aspects of radical reactions (see below).

(c) In a few cases, the transfer constant has been materially improved by employing a large excess of the addendum over the olefin. A convenient method to achieve this consists in adding the olefin very slowly to the addendum to ensure high concentration of the latter at all times. In this way, a yield of as high as 39% of the 1:1-adduct can be obtained from $CCl_4$ and vinyl acetate, from which, under normal conditions, only telomeric products would have resulted.

### (B) *Addition of aldehydes*

When catalysed by ultraviolet light or peroxides, aliphatic aldehydes, in particular straight chain aldehydes such as butanal or heptanal, add to olefins to give good yields of ketones, according to the following (Kharasch *et al.*):[9]

$$R'—CHO \rightarrow R'—\dot{C}O$$

$$R'—\dot{C}O + R—CH{=}CH_2 \rightarrow R—\dot{C}H—CH_2—COR'$$

$$R—\dot{C}H—CH_2—COR' + R'—CHO \rightarrow R—CH_2—CH_2COR' + R'—\dot{C}O$$

etc.

The radicals derived from benzaldehyde, and $\alpha\beta$ unsaturated aldehydes such as crotonaldehyde, however, do not undergo this reaction. The former, in fact, adds to benzaldehyde itself, addition taking place at the oxygen atom in the carbonyl function,[10] as follows:

$$\text{Ph—CHO} \rightarrow \text{Ph—}\dot{\text{C}}\text{=O}$$

$$\text{Ph—}\dot{\text{C}}\text{=O} + \text{Ph—CHO} \rightarrow \text{Ph—CO—O—}\dot{\text{C}}\text{H—Ph} \rightarrow [\text{Ph—CO—O—}\overset{|}{\text{C}}\text{H—Ph}]_2$$

The resulting radical then dimerizes to a mixture of the meso and racemic dimer.

It is worthy of note that aldehydes add particularly well to olefinic bonds conjugated with other unsaturated groups, as that in methyl crotonate. This finds an explanation in the fact that the process $\text{R—}\dot{\text{C}}\text{=O} \rightarrow \text{R—}\overset{+}{\text{C}}\text{=O} + e$, which leads to the formation of a resonance stabilized cation, proceeds with ease, and thus addition (of the acyl radical) to a conjugated, as compared with an isolated, double bond, would be facilitated by virtue of the 'electron sink' (such as the ester group in crotonate) through participation of polar transition states such as (6).

$$\begin{array}{ccc} \text{CH}_3\text{CH=CH—C(=O)OEt} & \leftrightarrow & \text{CH}_3\dot{\text{C}}\text{H—CH=C(O}^-\text{)OEt} \\ \text{R—}\dot{\text{C}}\text{=O} & & \vdots \\ & & \text{R—}\overset{+}{\text{C}}\text{=O} \end{array}$$

(6)

## (C) *Addition of ketones*

Ketones do not add homolytically to olefins generally. A known exception is cyclohexanone, which gives with 1-octene small yields of 2-octylcyclohexanone and chloroacetone, which adds to a number of olefins to give 1:1-adducts in modest yields (ca. 20%). It is noteworthy that chloroacetone adds as H and ClCH—CO—$CH_3$, the bond undergoing fission in the initiation and propagation steps being the C—H rather than the C—Cl bond, reminiscent of chloroform (Huang).[11]

(D) *Addition of esters*

Until recently the only esters found to add well to olefins had been the $\alpha$ brominated esters, such as ethyl bromoacetate, which with 1-octene gives the $\gamma$ bromo-ester (7) in good yield (Kharasch *et al.*).[12] The failure of

$$Br{-}CH_2{-}CO_2Et \rightarrow Br\cdot + \cdot CH_2{-}CO_2Et$$

$$R'{-}CH{=}CH_2 + \cdot CH_2CO_2Et \rightarrow R'{-}\dot{C}H{-}CH_2{-}CH_2CO_2Et$$

$$R'{-}\dot{C}H{-}CH_2{-}CH_2CO_2Et + Br{-}CH_2CO_2Et$$

$$\rightarrow R'{-}CHBr{-}CH_2{-}CH_2{-}CO_2Et + \cdot CH_2CO_2Et \quad (7)$$

$$(R' = C_6H_{13})$$

unsubstituted esters to add is no doubt due to an inefficient transfer step, which incurs the breaking of a C—H bond, a process much less facile than fission of the C—Br bond in the bromo-esters. With the conditions for chain transfer improved by employing a large excess of the addendum (20–50 moles per mole of olefin), Hey, Cadogan, and co-workers[13] have found it possible to realize fair yields of 1:1-adducts from 1-octene and the following substituted esters: methyl malonate, acetoacetate, and cyano-acetate.

Under similar conditions the addition to olefinic bonds in conjugated systems by ethyl cyanoacetate has afforded what may be considered a synthetic route complementary to the Michael reaction (Huang, Ong, and Ong).[6] Thus radical addition to $\alpha$ methylstyrene, methyl cinnamate, benzylidene-acetone and cinnamonitrile (8) gives the adducts (9) while under conditions of the Michael reaction (10) becomes the main product. Methyl crotonate, however, gives the same product (11) with the cyanoacetate under

$$Ph{-}CH{=}CH{-}X \ (8) \nearrow Ph{-}CH_2{-}CHX{-}CH(CN)CO_2Me \ (9) \quad \text{(radical addition)}$$

$$Ph{-}CH{=}CH{-}X \ (8) \searrow \begin{array}{l} PhCH{-}CH_2{-}X \\ \quad | \\ CH(CN){-}CO_2Me \end{array} \ (10) \quad \text{(Michael addition)}$$

$$(X = Me, CO_2Me, COMe, \text{ and } CN)$$

$$CH_3CH{=}CH{-}CO_2Me \rightarrow \begin{array}{l} CH_3CH{-}CH_2CO_2Me \\ \quad\ | \\ \quad CH(CN){-}CO_2Me \end{array} \quad (11)$$

either set of conditions. The orientation in the radical reaction can be explained in terms of the relative stabilizing influence of the substituent groups, namely Ph and X in (8) and $CH_3$ and $CO_2Me$ in methyl crotonate, on the intermediate radical.

### (E) *Addition of alkanes and arylalkanes*

Very few hydrocarbons are known to add to olefins effectively. Cyclohexane, under the modified conditions just described for the esters (i.e., using a large excess of the addendum), gives with 1-octene a 40% yield of 1-cyclohexyloctane, providing a convenient alternative to the Wurtz synthesis (Hey, Cadogan, and Ong).[14] Addition of toluene to 1-hexene or 1-decene affords only low yields of the 1:1-adducts. These are accompanied by much bibenzyl and telomers, and in each case a good deal of the olefin is recovered. The formation of bibenzyl, as well as the recovery of olefin, indicate a sluggish addition step, which is to be expected considering the relative inactivity of the resonance stabilized benzyl radical, while telomer formation points to inefficient chain transfer. However, when the olefinic bond is conjugated with an unsaturated group, addition appears to proceed much more readily. Thus methyl crotonate has been found to give no less than 59% of the 1:1-adduct with toluene (Huang *et al.*).[15] With $\alpha$ methoxytoluene as the addendum, the yield of 1:1-adduct is again considerably higher with a conjugated olefin like methyl crotonate than with simple olefins like 1-decene (Huang and Loke).[16] The same polar effect believed to operate in the addition of the acyl radical, derived from aldehydes, may be invoked here to explain this facilitation of addition, since the benzyl or substituted benzyl radical will, by virtue of its readiness to relinquish an electron to give the resonance stabilized benzyl cation, stabilize polar transition states such as (12).

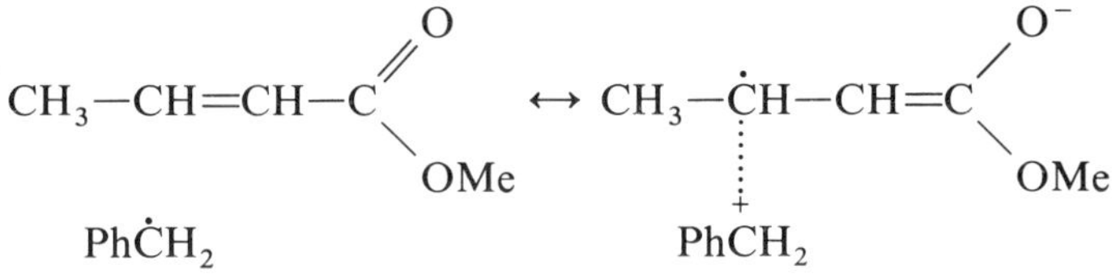

$$\underset{\mathrm{Ph\dot{C}H_2}}{\mathrm{CH_3{-}CH{=}CH{-}C({=}O)OMe}} \longleftrightarrow \underset{\overset{+}{\mathrm{PhCH_2}}}{\mathrm{CH_3{-}\dot{C}H{-}CH{=}C(O^-)OMe}}$$

(12)

### (F) *Addition of sulphur compounds*

A variety of sulphur compounds, such as thiols, hydrogen sulphide, bisulphite ions, sulphenyl halides, sulphur chloride pentafluoride, sulphonyl, and sulphuryl halides, add by a radical mechanism to olefinic and acetylenic bonds. On each of these classes of compounds extensive study has been conducted but as a typical example only the addition of thiols will be treated here.

The first example of the reaction of thiols with olefins, namely, the anti-Markownikov addition of thiophenol to styrene, was reported in 1905 and found to exhibit the characteristic marks of a radical chain process (effect of air and acceleration by light) by Ashworth and Burkhardt[17] twenty three years later. The currently accepted mechanism is the one formulated by Kharasch, Read, and Mayo[18] in 1938 and is exemplified by the methyl mercaptan-ethylene system:

$$CH_3{-}S\cdot + CH_2{=}CH_2 \rightarrow CH_3SCH_2\dot{C}H_2 \qquad (\Delta H = -14\,\text{kcal})$$

$$CH_3SCH_2\dot{C}H_2 + HSCH_3 \rightarrow CH_3SCH_2CH_3 + CH_3S\cdot \quad (\Delta H = -12\,\text{kcal})$$

Both addition and displacement are appreciably exothermic, resulting in a rapid reaction of long chain length, for which values as high as $10^4$ have been reported (Back, *et al.*).[19] There is ample evidence that in some cases the addition step is reversible (Sivetz, 1959).[20] The details of the mechanism are similar to those of other radical additions described earlier. It is relevant to mention that, under suitable conditions, thiols add 'normally' to olefins by an ionic mechanism, according to the Markownikov rule, whereas radical addition leads to the 'abnormal' product. However, $\alpha\beta$ unsaturated carbonyl compounds and nitriles lead to the same products by both mechanisms.

For convenience the addition reactions of carbenes are discussed separately (Chapter 17).

## 12.4 Stereochemistry of addition reactions

The preferred mode of addition by radicals has been shown to be *trans*. Skell[21] studied the addition of deuterium bromide to *cis* and *trans* but-2-enes at −80 °C, and from the fact that he obtained the *threo* product from the former and the *erythro* product from the latter concluded that D and Br had added *trans* to the double bond. Goering[22] investigated the addition of hydrogen bromide to the 2-bromobut-2-ene system, and as his findings also demonstrate the effect of variations in reaction conditions, we shall describe these briefly here.

At a low temperature (−80 °C) and in the presence of a large excess of hydrogen bromide, it is found that stereospecific addition results. Thus, the *cis* bromide gives the *meso* dibromide, whereas the *trans* bromide leads to the racemic modification, results indicative of *trans* addition to the olefinic bond. At a relatively high temperature (around room temperature) and a low concentration of the addend, however, *trans* addition apparently no longer holds, and either of the olefin is found to give the same mixture of meso and racemic products.

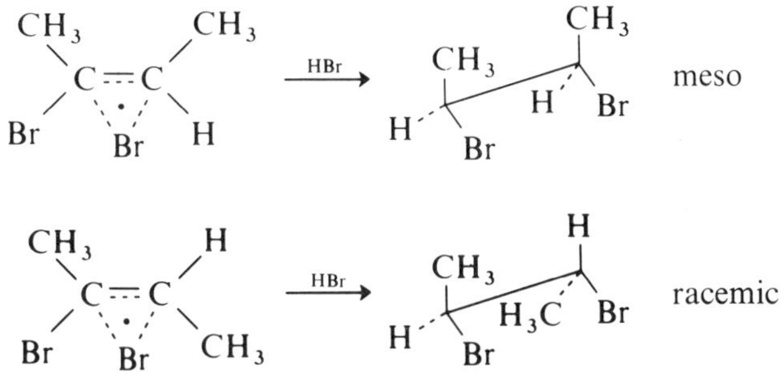

$CH_3$ $CH_3$ C=C Br Br H $\xrightarrow{HBr}$ $CH_3$ H Br $CH_3$ H Br meso

$CH_3$ H C=C Br Br $CH_3$ $\xrightarrow{HBr}$ $CH_3$ H Br H $H_3C$ Br racemic

The stereospecificity of this reaction has been explained in terms of 'bridged' radicals as shown in the diagrams, for which the transfer step (abstraction of a hydrogen atom from hydrogen bromide) would have to occur on the side remote from the 'bridge'. At higher temperatures one bridge structure can be envisaged as being interconvertible with the other via the open chain radicals, resulting in the same mixture of two possible conformations, before it has time to abstract a hydrogen atom from the addendum. Thus the same mixture of products is formed from either olefin at higher temperatures.

## 12.5 Orientation in radical addition reactions

The insensitivity of radical addition reactions towards polar effects has found experimental evidence (Haszeldine)[23] in the addition of atomic bromine and trichloromethyl radicals to the olefins $XCH{=}CH_2$, which takes place exclusively at the terminal carbon atom irrespective of widely different electronegativities of the substituent X (X = Me, Cl, F, COOMe, $CF_3$, and CN). More recently Cadogan[24] confirmed this point by showing that nuclear substituents have hardly any effect on the direction in which the trichloromethyl radical adds to the olefinic double bond in *trans* stilbene. The addition of a radical R· to an unsymmetrically substituted olefin XCH=CHY to give the radical $X\dot{C}H{-}CHRY$ or $XCHR{-}\dot{C}HY$ would thus be determined solely by the relative stabilizing influence of the substituents X and Y on the intermediate alkyl radical. Methyl crotonate, for instance, could give the radicals (13) or (14), leading in turn to isomeric adducts, depending on whether the methyl or the ester group, respectively, has the greater stabilizing power.

$$CH_3CH{=}CHCOOMe \xrightarrow{R\cdot} CH_3\dot{C}H{-}CHRCOOMe \quad (13)$$

$$CH_3CH{=}CHCOOMe \xrightarrow{R\cdot} CH_3CHR{-}\dot{C}HCOOMe \quad (14)$$

It would therefore be useful to establish a scale of relative stabilizing capacities of substituent groups known to confer stability in alkyl radicals, as such a scale would enable predictions to be made as to the orientation in addition reactions as well as the course of other reactions in which the stabilization factor plays a dominant role. This has been achieved. From

**Table 12.2** Addition of Bromotrichloromethane to X—CH═CH—Y

| X | Y | % Yield of $XCH(CCl_3)$—CHYBr |
|---|---|---|
| Me | $CO_2Et$ | 84 |
| $CO_2Et$ | Ph | 72 |
| CN | Ph | 56 |
| $COCH_3$ | Ph | 57 |

**Table 12.3** Addition of Butanal to X—CH═CH—Y

| X | Y | % Yield of XCH(COPr)—$CH_2Y$ |
|---|---|---|
| Me | $CO_2Et$ | 77 |
| Me | $CO_2H$ | 83 |
| $CO_2H$ | $CO_2Et$ | 49 |
| $CO_2Et$ | $COCH_3$ | 56 |

a systematic study (Huang)[25] of the addition of butanal and bromotrichloromethane to a series of unsymmetrically substituted olefins X—CH═CH—Y, including ethyl crotonate, crotonic acid, ethyl hydrogen maleate, ethyl $\beta$-acetoacrylate, ethyl cinnamate, and *cis* and *trans* cinnamonitrile (see Tables 12.2, 12.3), the following order of stabilizing capacities has been established:

$$Ph > CO \sim CN > CO_2R \sim CO_2H > CH_3 > H$$

**Table 12.4** Addition of Butanal and Heptanal (R—CHO) to $Me_2C{=}CH{-}X$

| X | Molar Ratio of 1:1-adducts $Me_2C(COR){-}CH_2X$ to $Me_2CH{-}CHX(COR)$ |
|---|---|
| COMe | 10:1 |
| $CO_2Et$ | 3:1 |
| $CO_2H$ | 3:1 |
| CN | ∞:1 |

Similarly, addition of butanal and heptanal to the olefins $Me_2C{=}CHX$ (X = COMe, $CO_2Et$, $CO_2H$, CN), in each of which the stabilizing influence of X is pitted against that of two methyl groups (Table 12.4), gives

$$CN > COMe > CO_2Et \sim CO_2H$$

By way of illustration the addition to ethyl cinnamate and to ethyl β-acetoacrylate are briefly described below.

Addition of bromotrichloromethane to ethyl cinnamate led to an adduct which could be either (15) or (16). Since dehydrobromination with potassium acetate, followed by hydrolysis with perchloric acid in acetic acid led to benzylidenemalonic acid, the adduct must have been (15), and hence in stabilizing effect, Ph > $CO_2Et$.

$$Ph{-}CH{=}CH{-}CO_2Et \nearrow Ph{-}\underset{}{\overset{Br}{CH}}{-}\overset{CCl_3}{CH}{-}CO_2Et\ (15) \rightarrow Ph{-}CH{=}C(CO_2H)_2$$

$$Ph{-}CH{=}CH{-}CO_2Et \searrow Ph{-}\overset{CCl_3}{CH}{-}\overset{Br}{CH}{-}CO_2Et\ (16)$$

Addition of butanal to ethyl β-acetoacrylate gave an adduct which could be either (17) or (18). Hydrolysis of the adduct with hot alcoholic potassium hydroxide was found to bring about decarboxylation also, to give the diketone (19). Hence the adduct must have been (17), as only the acid derived from this β-keto-ester, and not that from the isomeric (18), would be expected to be prone to decarboxylation on heating. Thus CO > $CO_2Et$.

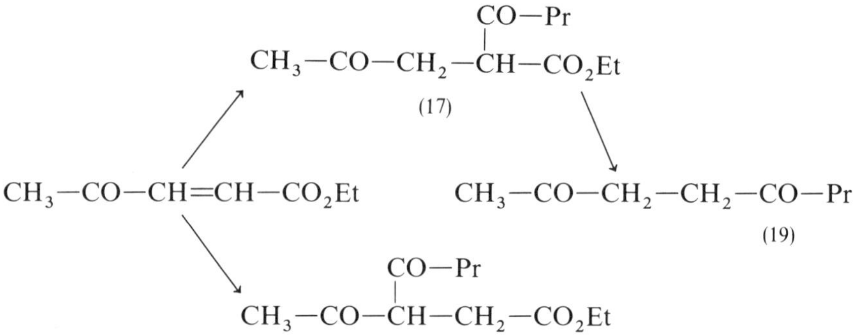

$$CH_3{-}CO{-}CH_2{-}CH(CO{-}Pr){-}CO_2Et \quad (17)$$

$$CH_3{-}CO{-}CH{=}CH{-}CO_2Et \qquad CH_3{-}CO{-}CH_2{-}CH_2{-}CO{-}Pr \quad (19)$$

$$CH_3{-}CO{-}CH(CO{-}Pr){-}CH_2{-}CO_2Et \quad (18)$$

## 12.6 Intramolecular addition reactions

Intramolecular addition reactions, which give rise to cyclic products, have yielded some interesting results. A stabilized radical such as (20) adds intramolecularly to form a cyclohexane derivative as expected but the primary hexa-5-enyl radical (21) gives a five-membered ring compound,[26] as shown. The preference for a six-membered ring in the first reaction is

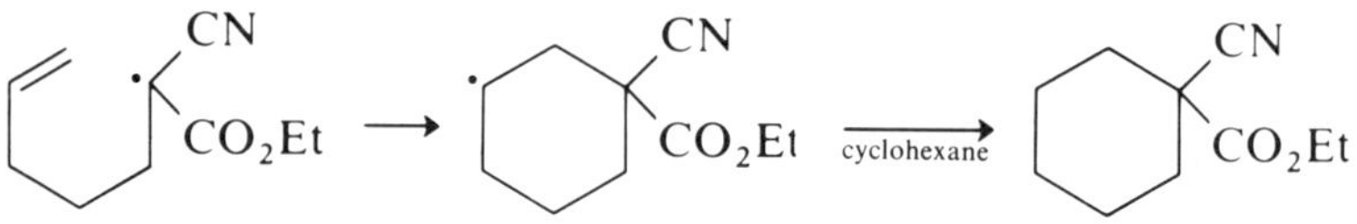

CN, $CO_2Et$ → CN, $CO_2Et$ $\xrightarrow{\text{cyclohexane}}$ CN, $CO_2Et$

(20)

$\cdot CH_2$ → $\dot{C}H_2$ $\xrightarrow{R_3SnH}$ $CH_3$

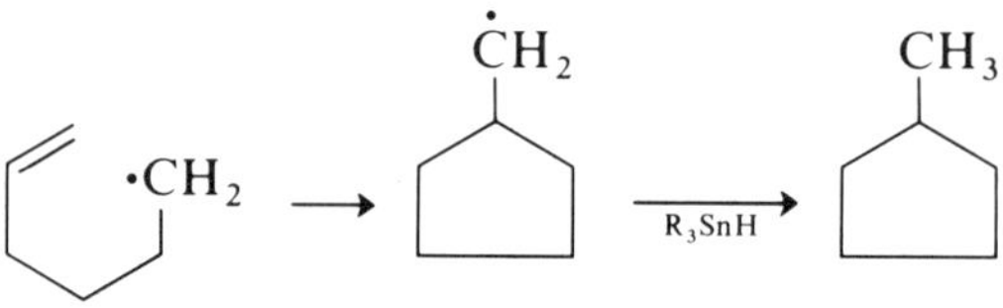

(21)

believed to be due to thermodynamic control resulting from the reversibility of addition of the resonance stabilized cyanoacetate radical. This is supported by a recent demonstration of reversibility in the homolytic addition of ethyl cyanoacetate to conjugated olefins.[27] The second reaction, on the other hand, is a fast process involving a relatively reactive radical, the rate constant[28] having been estimated to be ca. $10^5\ s^{-1}$, and its cyclization to a five-membered ring can thus be attributed to kinetic, rather than thermodynamic control. Cyclizations of other radicals similar to (21) have been reported.[29]

An interesting example[26b] of steric factors predominating over stabilizing effects may be found in the cyclization of the radical (22) in which addition takes place on the non-terminal carbon atom to give the six-membered ring (23), rather than on the terminal carbon atom to form the seven-membered homologue (24).

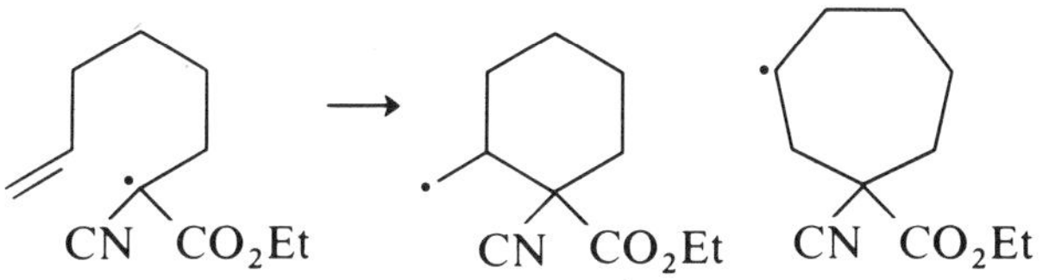

CN $CO_2Et$ → CN $CO_2Et$   CN $CO_2Et$

(22) (23) (24)

The cyclization reaction has found application in the synthesis of polycyclic compounds, for instance in the formation of the perhydrophenanthrene derivative (25) by double ring closure:[30]

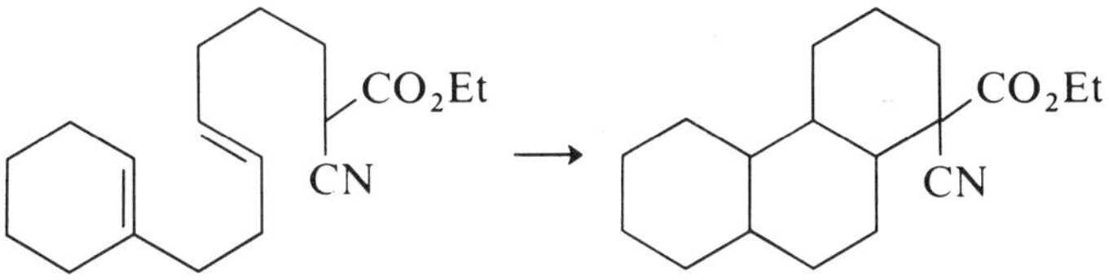

$CO_2Et$ CN → $CO_2Et$ CN

(25)

## 12.7 Radical polymerization[31]

### (A) *Kinetics*

In radical addition reactions it is noted that one of the side reactions is telomerization in which the intermediate radical $R'—\dot{C}HCH_2—Y$ is incorporated with a number of olefin molecules before undergoing chain transfer with the addendum. If this process is extended to a large number of olefins, a polymer results. Polymers produced by radical processes often have $10^3$ to $10^5$ repeated units per molecule and may have molecular weights as high as $10^5$ to $10^7$.

Radical polymerizations of ethylene and substituted ethylenes give a range of useful polymers. Well known examples are polyethylene, for containers, film, pipe, electrical insulation, etc.; polystyrene, used as foamed plastic, or for insulation and packing; poly(vinyl chloride), for pipes, insulation, shower curtains, leatherette, and hoses; poly(methylmethacrylate), for watch lenses, contact lenses, aircraft windows, etc.;

and polyacrylonitrile, for use as fibres. Thus vinyl polymerization is of great industrial importance.

The mechanism of polymerization can be considered in terms of the three main steps of initiation, propagation, and termination, as shown in the accompanying scheme.

$$\begin{array}{lrcl} \textit{Initiation:} & \text{Initiator} & \xrightarrow{k_d} & 2\text{R}\cdot \\ & \text{R}\cdot + \text{M} & \xrightarrow{k_i} & \text{M}\cdot \\ \textit{Propagation:} & \text{M}_n\cdot + \text{M} & \xrightarrow{k_p} & \text{M}\cdot_{n+1} \\ \textit{Termination:} & \text{M}_n + \text{M}_m & \xrightarrow{k_t} & \text{M}_{n+1} \\ & \text{M}_n + \text{M}_m & \longrightarrow & \text{M}_n + \text{M}_m \end{array}$$

M is a molecule of monomer and $M_n\cdot$ is a polymeric radical with n monomer units in the chain. If the reasonable assumption is made that the rate constant for the reaction is independent of the chain length of the polymeric radical taking part, then, at the steady state, the rate of formation of radicals = rate of destruction of radicals i.e.,

$$2k_d f[\text{In}] = 2k_t[\text{M}\cdot]^2$$

where $f$ = fraction of the primary radicals that escape cage recombination and are thus able to initiate polymerization, and In = Initiator.

Hence

$$[\text{M}\cdot] = \sqrt{\frac{k_d f[\text{In}]}{k_t}}$$

This is a useful expression for the concentration of radicals present in the reacting system. The rate of formation of polymer is then given by:

$$\frac{-d[\text{M}]}{dt} = R_p = k_i[\text{R}\cdot][\text{M}] + k_p[\text{M}\cdot][\text{M}]$$

Since the monomer molecules consumed in the initiation step are negligible, the rate of consumption of monomer can be approximated as

$$R_p = k_p[\text{M}\cdot][\text{M}]$$

Substituting for $[\text{M}\cdot]$, we get

$$R_p = \frac{k_p}{k_t^{\frac{1}{2}}}\{k_d f[\text{In}]\}^{\frac{1}{2}}[\text{M}]$$

This equation predicts that the rate of formation of polymer should be (a) proportional to the square root of the initiator concentration and (b) proportional to the concentration of monomer. The first factor has been borne out for a large number of initiators and the second factor has been found to be generally true.

The ratio of rate constants $k_p/k_t^{\frac{1}{2}}$ represents the polymerizability of the monomer and can be obtained very easily by measuring the rate of polymer formation using an initiator with a known rate of decomposition. A larger value of $k_p/k_t^{\frac{1}{2}}$ means that a given amount of initiator will produce more polymer in a given time. Table 12.5 gives the rate constants found for some common monomers.

**Table 12.5** Rate Constants for Some Common Monomers at 25 °C*

| | $k_p/k_t^{\frac{1}{2}}$ | $k_p$ | $k_t \times 10^{-7}$ |
|---|---|---|---|
| *p*-Methoxystyrene | 0.00284 | 2.9 | 0.10 |
| Styrene | 0.014 | 40 | 0.79 |
| Methyl methacrylate | 0.055 | 275 | 2.5 |
| Vinyl acetate | 0.18 | 1000 | 3.0 |
| Methyl acrylate | 0.20 | 1500 | 5.5 |

* Rate constants are in litre mole$^{-1}$ s$^{-1}$. G. M. Burnett, *Mechanism of Polymer Reactions*, Interscience Publishers, Inc., New York, 1954, pp. 223, 234.

Another useful term is the kinetic chain length $\nu$, which is defined as the average number of monomer molecules consumed by every radical which starts a polymerizing chain. This quantity is given by the ratio of $R_p$ to $R_i$ (or $R_t$) where $R_i$ and $R_t$ are the rates of initiation and termination, respectively:

$$\nu = R_p/R_i = R_p/R_t$$

Since

$$R_t = 2k_t[\text{M}\cdot]^2 \quad \text{and} \quad R_p = k_p[\text{M}\cdot][\text{M}]$$

$$\nu = k_p[\text{M}]/2k_t[\text{M}\cdot]$$

Eliminating radical concentrations,

$$\nu = k_p^2[\text{M}]^2/2k_tR_p$$

These expressions reveal (a) the inverse relationship between chain length and the reaction rate, $R_p$ and (b) that the chain length at constant reaction rate is proportional to the square of the concentration of the monomer involved.

The relationship between $\nu$ and polymer molecular weight (or the average degree of polymerization $\bar{P}$, see page 166) can be deduced from the definitions of these terms and from a knowledge of the relative importance of disproportionation and combination as modes of chain termination. If the polymer terminates by disproportionation, then $\bar{P}$ equals $v$ and $\bar{P}$ equals $2v$ for termination by combination.

The chain lengths attained before termination may be very high, typical polymers having some $10^5$ monomer units in the chain, as estimated from molecular weight data obtained by viscometry, osmosis, or ultracentrifugation.

### (B) *Copolymerization*

The copolymerization of two different monomers, such as styrene and butadiene, has produced a variety of useful plastics. This subject, accordingly, has acquired great importance in technology. When two monomers, $M_1$ and $M_2$, copolymerize, the following reactions can occur:

| | Rate |
|---|---|
| $M_1\cdot + M_1 \xrightarrow{k_{11}} M_1\cdot$ | $k_{11}[M_1\cdot][M_1]$ |
| $M_1\cdot + M_2 \xrightarrow{k_{12}} M_2\cdot$ | $k_{12}[M_1\cdot][M_2]$ |
| $M_2\cdot + M_1 \xrightarrow{k_{21}} M_1\cdot$ | $k_{21}[M_2\cdot][M_1]$ |
| $M_2\cdot + M_2 \xrightarrow{k_{22}} M_2\cdot$ | $k_{22}[M_2\cdot][M_2]$ |

where $M_1\cdot$ and $M_2\cdot$ are the polymeric radicals having monomer 1 or monomer 2 on the end of the chain. Assuming that (a) the reactivity of the polymeric radical depends only on the end monomer and not on the composition of the rest of the chain and (b) that the steady state in both $M_1\cdot$ and $M_2\cdot$ can be attained and $k_{11}$ and $k_{22}$ are regenerative steps, we have

$$k_{21}[M_2\cdot][M_1] = k_{12}[M_1\cdot][M_2]$$

The rate of disappearance of the two monomers is given by

$$\frac{-d[M_1]}{dt} = k_{11}[M_1\cdot][M_1] + k_{21}[M_2\cdot][M_1]$$

$$\frac{-d[M_2]}{dt} = k_{12}[M_1\cdot][M_2] + k_{22}[M_2\cdot][M_2]$$

$$\frac{d[M_1]}{d[M_2]} = \frac{[M_1]}{[M_2]}\left\{\frac{k_{11}[M_1\cdot] + k_{21}[M_2\cdot]}{k_{12}[M_1\cdot] + k_{22}[M_2\cdot]}\right\}$$

$$[M_1\cdot] = \frac{k_{21}}{k_{12}}\frac{[M_1]}{[M_2]} \times [M_2\cdot]$$

$$\frac{d[\mathrm{M}_1]}{d[\mathrm{M}_2]} = \frac{[\mathrm{M}_1]}{[\mathrm{M}_2]}\left\{\frac{\dfrac{k_{11}k_{21}}{k_{12}}\dfrac{[\mathrm{M}_1]}{[\mathrm{M}_2]} + k_{21}}{\dfrac{k_{12}k_{21}}{k_{12}}\dfrac{[\mathrm{M}_1]}{[\mathrm{M}_2]} + k_{22}}\right\}$$

$$= \frac{[\mathrm{M}_1]}{[\mathrm{M}_2]}\left\{\frac{k_{11}k_{21}[\mathrm{M}_1] + k_{12}k_{21}[\mathrm{M}_2]}{k_{12}k_{21}[\mathrm{M}_1] + k_{12}k_{22}[\mathrm{M}_2]}\right\}$$

$$= \frac{[\mathrm{M}_1]}{[\mathrm{M}_2]}\frac{\dfrac{k_{11}}{k_{12}}[\mathrm{M}_1] + [\mathrm{M}_2]}{[\mathrm{M}_1] + \dfrac{k_{22}}{k_{21}}[\mathrm{M}_2]}$$

$$= \frac{[\mathrm{M}_1]}{[\mathrm{M}_2]}\left\{\frac{r_1[\mathrm{M}_1] + [\mathrm{M}_2]}{[\mathrm{M}_1] + r_2[\mathrm{M}_2]}\right\}$$

where

$$r_1 = \frac{k_{11}}{k_{12}} \quad \text{and} \quad r_2 = \frac{k_{22}}{k_{21}}$$

This important relation gives the relative rates of incorporation of the two monomers into the polymer as a function of their concentration and $r_1$ and $r_2$.

The rate ratios $r_1$ and $r_2$ are known as monomer reactivity ratios. For $r_1 > 1$, the polymeric radical ending with monomer 1 prefers to react with its own monomer rather than with monomer 2; whereas for $r_1 < 1$, the polymeric radical ending in monomer 1 prefers to react with monomer 2. Under the special condition where $r_1 = r_2 = 1$, the polymeric radical has no preference for either monomer, irrespective of which monomer is on the end of the chain. In that case, the two monomers are distributed randomly along the polymer chain in a ratio depending only on the relative monomer concentrations in the reaction mixture. If $r_1$ and $r_2$ are both very small, the polymer will contain alternating units of monomer 1 and 2. Table 12.6 gives copolymer reactivity ratios found for some pairs of monomers.

### (C) *Transfer in polymerization*

An observation made in 1937 that polystyrene prepared in carbon tetrachloride had a relatively low molecular weight has been rationalized by Mayo[32] in terms of a chain transfer process, in which the growing polymeric

**Table 12.6** Copolymer Reactivity Ratios at 60 °C*

| $M_1$ | $M_2$ | $r_1$ | $r_2$ |
|---|---|---|---|
| Styrene | Methyl methacrylate | 0.5 | 0.5 |
| Styrene | *p*-Nitrostyrene | 0.2 | 1.1 |
| Styrene | *p*-Chlorostyrene | 0.7 | 1.0 |
| Styrene | *p*-Methoxystyrene | 1.2 | 0.8 |
| Styrene | Acrylonitrile | 0.4 | 0.04 |
| Styrene | Vinyl acetate | 55 | 0.01 |
| Styrene | Ethyl vinyl ether | 90 | ca. 0 |
| Styrene | Butadiene | 0.78 | 1.4 |
| Styrene | Vinyl chloride | 17 | 0.02 |
| Vinyl acetate | Vinyl chloride | 0.2 | 1.7 |
| Vinyl acetate | Methyl methacrylate | 0.01 | 20 |

* L. J. Young, *J. Polymer Sci.*, 1961, **54**, 411.

radical is envisaged as undergoing a transfer reaction with the solvent XY, thus terminating one chain and starting another, as shown below:

$$M_n\cdot + XY \xrightarrow{k_{tr}} M_n{-}X + Y\cdot$$

$$Y\cdot + M \xrightarrow{k_a} M\cdot$$

In the simplest case, the rate constant for the addition of Y· to the monomer is very large, and no retardation of polymerization results from transfer. In transfer reactions, therefore, the molecular weight of the polymer is changed without a change in the rate of polymerization. One measure of molecular weight is the degree of polymerization, defined as

$$\bar{P} = \frac{\text{Monomer units used}}{\text{Polymer molecules formed}}$$

In the absence of a transfer agent and assuming that displacement occurs without retardation and that termination is entirely by combination, we can derive the expression for the degree of polymerization ($\bar{P}_0$) in the absence of a solvent as follows:

$$M\cdot + M \xrightarrow{k_p} M\cdot$$

$$2M\cdot \xrightarrow{k_t} \text{Polymer}$$

$$\bar{P}_0 = \frac{k_p[M\cdot][M]}{k_t[M\cdot]^2} = \frac{k_p[M]}{k_t[M\cdot]}$$

In the presence of a transfer agent however, we have seen that there is another way by which the polymer molecules can be formed and taking this into account, the expression for $\bar{P}$ becomes

$$\bar{P} = \frac{k_p[M\cdot][M]}{k_t[M\cdot]^2 + k_{tr}[M\cdot][XY]}$$

$$\frac{1}{\bar{P}} = \frac{k_t[M\cdot]}{k_p[M]} + \frac{k_{tr}}{k_p}\cdot\frac{[XY]}{[M]}$$

$$= \frac{1}{\bar{P}_0} + C\frac{[XY]}{[M]}$$

where $C = k_{tr}/k_p$ a transfer constant for the particular transfer agent XY. By studying the polymerization of the same monomer, e.g., styrene, in a number of solvents, the transfer constant for the solvents can be readily determined. In this case, the transfer equation becomes

$$\frac{1}{\bar{P}} = \text{constant} + C\frac{[XY]}{[M]}$$

and a plot of the inverse of the degree of polymerization against the mole ratio of the transfer agent to monomer should give a straight line, with a slope equal to the transfer constant. Table 12.7 lists some typical results, and values of $C$ for a number of solvents.

A low transfer constant indicates that the radical reaction between the growing styrene chain and the solvent is negligible compared with the monomer. A value of about 0.01, as determined for carbon tetrachloride

**Table 12.7** Transfer Constant for Styrene with some Transfer Agents* (all $\times 10^4$)

| Transfer Agent | $C_{60}$ |
|---|---|
| Cyclohexane | 0.024 |
| Benzene | 0.018 |
| Chloroform | 0.5 |
| Carbon tetrachloride | 90 |
| n-Butylmercaptan | $22 \times 10^4$ |
| n-Dodecylmercaptan | $19 \times 10^4$ |
| Ethyl thioglycolate | $58 \times 10^4$ |

* From C. Walling, *Free Radicals in Solution*, John Wiley and Sons Inc., New York, 1957, p. 241.

at 60 °C, means that the polymerizing free radical attacks carbon tetrachloride about one hundredth as readily as it adds another monomer unit to the chain. On the other hand, the high values for the mercaptans and ethyl thioglycolate indicate the relatively high reactivity of these transfer agents. These compounds are commonly known as regulators or modifiers and find considerable application as such. An important example is found in the manufacture of synthetic rubber, in which thiols are commonly included in the polymerization formulas in order to control the molecular weight of the polymer.

Chain transfer may occur with solvents as well as with polymer and monomer. The probability of transfer reaction becomes more important as the size of the polymer radical increases, giving rise to the formation of branches along the polymer chain. In this process hydrogen is removed from the polymer molecule and the radical formed starts another chain, resulting in the formation of a branch at the radical centre (as shown below). This reaction is very important in the polymerization of ethylene, due to its resultant effect on the mechanical properties of the polymer. Many of the branches along the polythene molecule are short, consisting of four carbon atoms suggesting that these might arise by intramolecular chain transfer.

$$\begin{array}{c}\sim CH_2\\ H_2C\cdot \end{array} \longrightarrow \begin{array}{c}\sim CH\cdot\\ H_3C \end{array} \longrightarrow \begin{array}{c}\sim CH \sim\\ H_3C \end{array}$$

Transfer to monomer occurs more readily in a system where the polymer radical is reactive and where the monomer has easily abstractable atoms such as those at the allylic positions. A typical example is illustrated below:

$$\sim CH_2-\underset{CH_2Cl}{\underset{|}{CH}}\cdot + \underset{CH=CH_2}{\underset{|}{CH_2Cl}} \rightarrow \sim CH_2\underset{CH_2Cl}{\underset{|}{CH_2}} + \overbrace{CH_2-CH-CHCl}^{\cdot}$$

## 12.8 Cycloaddition reactions

Another group of addition processes which are quite widespread are the reactions known as cycloadditions, in which interest has grown considerably since the elucidation of orbital symmetry requirements in chemical reactions by Woodward and Hoffmann (see Section 3.7). In this section we shall be concerned only with [2 + 2]-cycloadditions as these reactions involve radical, or more exactly, diradical intermediates. Common examples are of two main types, the reactions of fluorinated olefins or other suitably substituted olefins and the dimerization of allenes, as illustrated below:

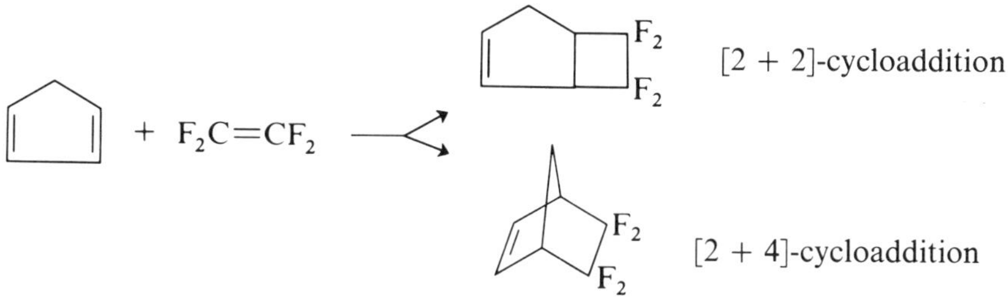

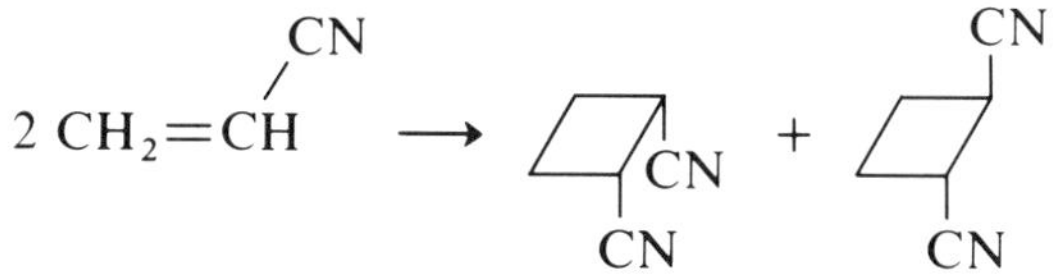

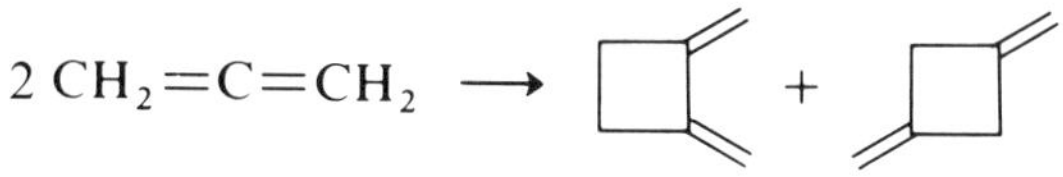

[2 + 2]-cycloaddition

+ $F_2C{=}CF_2$

[2 + 4]-cycloaddition

$2\,CH_2{=}CH(CN) \longrightarrow$ [cyclobutane with CN, CN] + [cyclobutane with CN, CN]

$2\,CH_2{=}C{=}CH_2 \longrightarrow$ [dimethylenecyclobutane] + [dimethylenecyclobutane]

The reaction of tetrafluoroethylene with cyclopentadiene gives both [2 + 4]- and [2 + 2]-cycloaddition products. It may be noted that [2 + 4]-cycloaddition, the well known Diels-Alder reaction, is a symmetry allowed reaction while [2 + 2]-cycloaddition is symmetry forbidden. Although the latter is symmetry forbidden it does not mean that the reaction cannot occur; it can in fact take place, not via one step concerted transition states but through formation of radical intermediates, usually diradicals. The formation of *cis* and *trans* 1,2-dicyanocyclobutanes from the dimerization of acrylonitrile is indicative of this non-concerted mechanism.

## References


1 G. Sosnovsky, *Free Radical Reactions in Preparative Organic Chemistry*, Macmillan, London, 1964.
2 W. A. Waters, Ed., *Vistas in Free Radical Chemistry*, Pergamon, London, 1959, p. 139.
3 C. Walling, *Free Radicals in Solution*, John Wiley and Sons Inc., New York, 1957, p. 241.
4 J. K. Kochi, *J. Am. chem. Soc.*, 1956, **78**, 4815.
5 J. M. Tedder, *Proc. chem. Soc.*, 1961, 80.
6 R. L. Huang, C. O. Ong, and S. H. Ong, *J. chem. Soc.* (*C*), 1968, 2217.
7 W. G. Bentrude and K. R. Darnall, *J. Am. chem. Soc.*, 1968, **90**, 3588.
8 M. S. Kharasch, E. V. Jensen, and W. H. Urry, *J. Am. chem. Soc.*, 1947, **69**, 1100.
9 M. S. Kharasch, B. M. Kuderna, and W. H. Urry, *J. org. Chem.*, 1949, **14**, 248.
10 F. F. Rust, F. H. Seubold, and W. E. Vaughan, *J. Am. chem. Soc.*, 1948, **70**, 3258.

11 R. L. HUANG, *J. chem. Soc.*, 1957, 2528.
12 M. S. KHARASCH, P. S. SKELL, and P. FISHNER, *J. Am. chem. Soc.*, 1948, **70**, 1055.
13 J. C. ALLEN, J. I. G. CADOGAN, B. W. HARRIS, and D. H. HEY, *J. chem. Soc.*, 1962, 4468.
14 D. H. HEY, J. I. G. CADOGAN, and S. H. ONG, *J. chem. Soc.*, 1965, 1939.
15 R. L. HUANG, H. H. LEE, and L. Y. WONG, *J. chem. Soc.*, 1965, 6730.
16 R. L. HUANG and S. E. LOKE, unpublished results.
17 F. ASHWORTH and G. N. BURKHARDT, *J. chem. Soc.*, 1928, 1791.
18 M. S. KHARASCH, A. T. READ, and F. R. MAYO, *Chemy. Ind.*, 1938, **57**, 752.
19 R. BACK, G. TRICK, C. MCDONALD, and C. SIVETZ, *Can. J. Chem.*, 1954, **32**, 1708.
20 C. SIVETZ, *J. phys. Chem.*, 1959, **63**, 34.
21 P. S. SKELL and R. G. ALLEN, *J. Am. chem. Soc.*, 1959, **81**, 5383.
22 H. L. GOERING and D. W. LARSEN, *J. Am. chem. Soc.*, 1959, **81**, 5937.
23 R. N. HASZELDINE, *J. chem. Soc.*, 1953, 1199.
24 J. I. G. CADOGAN and P. W. INWARD, *J. chem. Soc.*, 1962, 4169.
25 R. L. HUANG, *J. chem. Soc.*, 1956, 1749; 1957, 1342.
26*a* J. I. G. CADOGAN, D. H. HEY, and S. H. ONG, *J. chem. Soc.*, 1965, 1932;
*b* M. JULIA, Free Radical Cyclisation in *International Symposium on Free Radicals in Solution*, Butterworth, London, 1966, p. 167.
27 R. L. HUANG, C. O. ONG, and S. H. ONG, *J. chem. Soc.* (*C*), 1968, 2217.
28 D. J. CARLSSON and K. U. INGOLD, *J. Am. chem. Soc.*, 1968, **90**, 7047; J. F. GARST and F. E. BARTON, *Tetrahedron Lett.*, 1969, 587.
29 J. I. G. CADOGAN, M. GRUNBAUM, D. H. HEY, S. H. ONG, and J. T. SHARP, *Chemy. Ind.*, 1968, 422.
30 M. JULIA and F. LE GOFFIC, *Bull. Soc. chim. Fr.*, 1964, 1129; for a recent review see M. JULIA, *Acc. Chem. Res.*, 1971, **4**, 388.
31 For further reading on radical polymerization refer to the following books:
(i) ALFREY, BOHRER, and MARK, *Copolymerization*, Interscience, 1952.
(ii) BANFORD, BARB, JENKINS, and ONYON, *The Kinetics of Vinyl Polymerization by Radical Mechanisms*, Butterworth, London, 1958,
(iii) BEVINGTON, *Radical Polymerization*, Academic Press, 1961,
(iv) BILLMEYER, *Textbook of Polymer Science*, Interscience, New York, 1962,
(v) See reference 3.
32 F. R. MAYO, *J. Am. chem. Soc.*, 1943, **65**, 2324.

# 13
# Radical Aromatic Substitution

Known reactions under this heading have been confined mainly to the displacement, by an aryl, alkyl, or acyl radical R, of a hydrogen atom attached to an aromatic nucleus, as follows:

$$R + ArH \xrightarrow{-H\cdot} ArR$$

Among these reactions aromatic arylation has received by far the most intensive investigation, partly because of its utility in the synthesis of biaryls and certain polycyclic compounds, and partly because of the scope it offers for study of the mechanism of radical attack on the benzene nucleus. The major part of this Chapter will accordingly be devoted to this subject.

Substitution may be intermolecular or intramolecular, the latter being in fact an extension of the former, but because of its diverse applications in the synthesis of cyclic compounds is conveniently dealt with under a separate heading.

Historically the earliest instance of the direct homolytic arylation of an aromatic compound is probably that reported by Gomberg in 1924, although its significance in representing an entirely new type of aromatic substitution was completely overlooked for ten years afterwards. It remained for Grieve and Hey[1] in 1934 to draw attention to the fact that the results reported by Gomberg could not be fitted into the established pattern of either electrophilic or nucleophilic aromatic substitution but could be accommodated by invoking the agency of free radicals as the attacking entities. Since then a large number of reactions of a similar nature have been brought to light.[2]

## 13.1 Aromatic arylation as a synthetic method for biaryls

Arylation reactions may be classified according to the source of the arylating radical employed, as follows:

(a) reactions involving diazo, azo, and related compounds,
(b) reactions using aroyl peroxides and other sources of aroyloxy radicals,
(c) photochemical reactions, and
(d) miscellaneous reactions.

These are discussed briefly in the following paragraphs.

(A) *Diazo, azo, and related compounds*

The experimental technique of Gomberg's method consists in adding an aqueous solution of sodium hydroxide to a cold concentrated solution of a diazonium salt, covered with the aromatic compound to be arylated (e.g., benzene, chlorobenzene), with vigorous stirring. When the solution becomes alkaline the diazoate anion formed decomposes to give aryl radicals which arylate the aromatic compound present.

$$ArN_2^+Cl^- \longrightarrow Ar{-}N{=}N{-}OH \xrightarrow{PhH} Ph{-}Ar + N_2 + H_2O$$

Yields are generally low (5 to 40%) and mixtures of products are usually encountered. Numerous modifications of the method have been worked out, leading to improved yields and cleaner products: e.g., heating aniline with butyl and pentyl nitrite in the presence of benzene[3] gives a 65% yield of biphenyl.

Arylation in a homogeneous medium is possible by making use of acylarylnitrosamines,[4] as for example by the decomposition of nitrosoacetanilide in benzene, which proceeds smoothly around room temperature to give biphenyl, the initial step being the rearrangement of the nitroso compound to the diazoacetate, the source of the phenyl radical, as follows:

$$Ph{-}N(NO)COMe \longrightarrow Ph{-}N{=}N{-}OCOMe$$

$$\xrightarrow{PhH} Ph{-}Ph + N_2 + MeCO_2H$$

As to the manner in which the phenyl radicals are generated, the earlier notion that these are derived directly from the diazo compound, simultaneously with acetoxy radicals as indicated,

$$Ph{-}N{=}N{-}OCOMe \rightarrow Ph\cdot + N_2 + MeCO_2\cdot$$

has been invalidated by evidence of the intermediate formation of the diazo anhydride $PhN_2{-}O{-}N_2Ph$, which yields on homolysis phenyl and diazoate radicals, and which is considered to result from reaction between the diazonium acetate ion pair (with which the diazoacetate is known to be in equilibrium) and a diazoate anion, as shown below:

$$PhN{=}NOCOMe \rightleftarrows PhN_2^+MeCO_2^-$$

$$PhN{=}NOCOMe + MeCO_2^- \rightarrow (MeCO)_2O + Ph{-}N{=}NO^-$$

$$PhN_2^+MeCO_2^- + Ph{-}N{=}NO^- \longrightarrow PhN_2{-}O{-}N_2Ph$$

$$PhN_2{-}O{-}N_2Ph \rightarrow Ph\cdot + N_2 + PhN{=}NO\cdot$$

Decomposition of phenylazotriphenylmethane in boiling benzene gives, besides biphenyl, tetraphenylmethane and triphenylmethyl peroxide. That the benzene has taken part in the reaction is shown by the fact that, when it is replaced by chlorobenzene, the product becomes 4-chlorobiphenyl and not biphenyl. The arylating agent is again the phenyl radical, generated from the azo compound as follows:

$$PhN{=}NCPh_3 \rightarrow Ph\cdot + N_2 + \cdot CPh_3$$

However, in certain cases the triphenylmethyl radical tends to cause the occurrence of side reactions, as for example in the phenylation of nitrobenzene, in which this radical reacts with the nitro group in the substrate thereby considerably reducing the yield of nitrobiphenyls.

Other sources of arylating radicals include the triazens and aryl hydrazines, which give rise to aryl radicals as follows:

$$Ph{-}N{=}N{-}NHR \xrightarrow{\text{heat}} Ph\cdot + N_2 + \cdot NHR$$

$$Ph{-}NH{-}NH_2 \xrightarrow{Ag_2O} Ph{-}N{=}N\cdot \longrightarrow Ph\cdot + N_2$$

(B) *Aroyl peroxides and other sources of aryloxy radicals*

Decomposition of aroyl peroxides in aromatic substrates has probably become the most useful method of arylation, and one through which much of the fundamental work on mechanism has been carried out. A good deal of the earlier work was due to Gelissen and Hermans,[5] although the recognition of the process as one of homolytic substitution came later, arising from studies made on substituted benzenes by Hey and Wieland and their co-workers.[6] More recently it has been shown that the yields obtained in this reaction are significantly increased in the presence of oxygen[7] or catalytic amounts of certain nitrogenous compounds, such as nitrobenzene, nitrosobenzene, phenyl, and diphenylhydroxylamine.[8]

Arylation may also be carried out with lead tetrabenzoate which decomposes at about 125 °C to give aryl radicals as follows:

$$Pb(OCOPh)_4 \rightarrow Pb(OCOPh)_2 + 2PhCO_2\cdot$$

$$PhCO_2\cdot \rightarrow Ph\cdot + CO_2$$

or with phenyl iodosobenzoate, or the benzoate ion (by electrolysis) which gives benzoyloxy, and thence phenyl radicals as follows:

$$PhI(OCOPh)_2 \rightarrow PhI + 2PhCO_2\cdot$$

$$PhCO_2^- \rightarrow PhCO_2\cdot + e$$

Aryl radicals are also formed when aromatic carboxylic acids are oxidized with a hot solution of potassium persulphate.

### (C) *Photochemical reactions*

While photochemical reactions in general have been covered in Chapter 7, such processes as have applications in homolytic arylation have been deferred to this Chapter. These include the photolysis of aryl halides, in particular the iodides, and that of organometallic compounds.

The photolysis of aryl iodides results in the formation of aryl radicals which could react with the substrate. If the substrate is an aromatic solvent, biaryls are formed through this photochemical reaction which has been developed into a useful preparative method. For example, the photolysis of iodobenzene in cumene (isopropylbenzene) yields a mixture of isomeric isopropylbiphenyls and 2,3-dimethyl-2,3-diphenylbutane. This reaction can be extended to iodophenols and iodobenzoic acids; 2,4,6-tri-iodophenol and benzene react to give 2,4,6-triphenylphenol in 75% yield. *p*-Terphenyl could be obtained in 91% yield from 4-iodobiphenyl and benzene. Biphenyl is also obtained in this reaction. The relative yields of *p*-terphenyl and biphenyl are influenced by the presence of oxygen which favours the formation of the phenylation product, *p*-terphenyl over that of the hydrogen transfer product, biphenyl. The homolytic nature of these reactions has been inferred from the study of isomer distribution ratios of the products.

Another method of obtaining aryl radicals is the photolysis of organometallic compounds. Diphenylmercury or tetraphenyllead and cumene give a mixture of isomeric isopropylbiphenyls and the dimer, 2,3-dimethyl-2,3-diphenylbutane as well as the metal. It has been found that the ratio of isomeric biphenyls thus obtained differs slightly from that determined in the phenylation reaction using iodobenzene or benzoyl peroxide. Phenylation of toluene, *t*-butylbenzene and pyridine via a free radical mechanism could be carried out by photolysis of triphenylbismuth in the presence of the aromatic solvent.

### (D) *Miscellaneous reactions*

The formation of free radicals in reactions of Grignard reagents in the presence of catalytic quantities of cobaltous halides was postulated many years ago by Kharasch and his co-workers.[9] This was further substantiated by experiments in which the production of aryl radicals by this method was attempted in the presence of a suitable aromatic solvent. For instance, the reaction of *p*-bromotoluene with methylmagnesium iodide in the presence of cobaltous chloride and an excess of benzene produced 4-methylbiphenyl, showing participation of the *p*-tolyl radical formed from *p*-bromotoluene. Similarly, reactions with monosubstituted benzenes have been carried out and an analysis of the products showed the usual isomer distribution ratios for homolytic substitution, but with a decrease in the yield of the *ortho* isomer, which has been attributed to steric factors.[10]

When it is desired to effect substitution on a solid aromatic compound phenylation can be carried out at elevated temperatures, in the molten compound as solvent. For such reactions diazoaminobenzene may be used as the radical source, but a class of reagents of wider application is the aromatic sulphonyl halides, typical of which is benzenesulphonyl chloride which undergoes homolysis at about 255 °C with liberation of phenyl radicals. Biphenyl has thus been phenylated to give a 75 % yield of a mixture of *ortho*, *meta*, and *para* terphenyls in proportions 47, 28, and 25 %, respectively. Reaction with naphthalene results in an 80 % yield of a mixture of 1- and 2-phenylnaphthalene, in the ratio of four to one. These results closely resemble those obtained from the reaction of naphthalene with benzoyl peroxide. Similar experiments have been reported with disulphonyl chlorides and with sulphonyl bromides, while naphthalene 1- and 2-sulphonyl chlorides with naphthalene give the corresponding binaphthyls.[11]

## 13.2 Factors influencing the course of aromatic arylation

Early work revealed two striking characteristics in radical aromatic substitution reactions. These are (a) the independence of orientation on the electronic effects of nuclear substituents present in the aromatic compound arylated, and (b) the overall, albeit small, activating effect of all such substituents. Thus, be these *ortho*, *meta*, or *para* directing in heterolytic substitution reactions all were found to direct the incoming group mainly to the *ortho* and *para* positions, whilst even the nitro group was shown to have an activating effect. This, however, has since been found to be an oversimplified picture. Later findings, made possible through the more refined analytical techniques of infrared spectroscopy and gas chromatography, have brought about a clearer understanding of the mechanistic paths, in connection with which the reaction of three classes of compounds, in particular, have received detailed study. These are the peroxides, the acylarylnitrosamines, and the phenylazotriphenylmethanes. Among these the peroxides, especially benzoyl peroxide, have received by far the greatest attention, and their reactions will be described below in some detail.

The reactions of benzoyl peroxide with benzene derivatives have been studied both kinetically, and by the method of Partial Rate Factors which in the elucidation of the mechanism of electrophilic aromatic substitution reaction more than thirty years ago had proved to be of outstanding value.

The application of this latter method has shed much light on the factors influencing the course of the reaction, and will be briefly discussed here. Results from kinetics studies will be dealt with later.

The partial rate factors $F_o$, $F_m$, and $F_p$ are numerical values given for reactivities at the *ortho*, *meta*, and *para* positions, respectively, in the monosubstituted benzene PhR, relative to the reactivity at any one position in the unsubstituted benzene. These factors are calculated from experimentally determined values obtained from the measurement, by analysis of the substitution products, of (a) the ratio of the total substitution rate in PhR to that in benzene and (b) the relative yields of the *ortho*, *meta*, and *para* isomers. The former quantity is arrived at through a competition reaction between the two substrates, PhR and PhH, both present in large excess, and determining the extent to which the group R is present in the substitution product; while the latter is obtained through spectroscopic methods. More recently vapour phase chromatography has proved a direct, and more accurate, analytical method for estimation of these quantities. The results[2,27] obtained for the phenylation of a series of monosubstituted benzenes with benzoyl peroxide at 80 °C are given in Table 13.1, and from these the following striking features have been noted:

**Table 13.1** Relative Rates and Partial Rate Factors for the Phenylation of Benzene Derivatives with Benzoyl Peroxide at 80 °C*

| R in PhR | Rate Ratio (PhH = 1) | Partial Rate Factors $F_o$ | $F_m$ | $F_p$ |
|---|---|---|---|---|
| $NO_2$ | 2.9 | 5.5 | 0.86 | 4.9 |
| F | 1.0 | 1.7 | 0.95 | 0.86 |
| Cl | 1.1 | 3.65 | 1.55 | 1.98 |
| Br | 1.3 | 2.95 | 1.64 | 1.78 |
| I | 1.3 | 2.0 | 1.3 | 1.3 |
| Me | 1.2 | 2.68 | 0.70 | 2.02 |
| Et | 0.90 | 1.4 | 0.76 | 1.0 |
| *i*-Pr | 0.64 | 0.60 | 0.81 | 1.0 |
| *t*-Bu | 0.64 | 0.73 | 1.52 | 1.63 |
| CN | 3.7 | 6.5 | 1.1 | 6.5 |
| $CO_2Me$ | 1.8 | 3.0 | 0.93 | 2.7 |
| Ph | 2.9 | 2.1 | 1.0 | 2.5 |

* Data from D. H. Hey, *Advances in Free-Radical Chemistry*, 1967, **2**, G. H. Williams, Ed., Logos, London.

(a) Neither the total rates nor the partial rate factors differ greatly from unity, in contrast to the relatively much larger values usually encountered in ionic substitution reactions.

(b) Irrespective of their polar characters all substituents, except for the isopropyl and *t*-butyl groups, activate the benzene nucleus towards phenylation, albeit to a much smaller extent compared with ionic substitution reactions. Furthermore, the nitro and cyano groups, which in heterolytic aromatic substitution have a strong depressing effect on the nucleus, are seen to be, among the substituents listed, the most strongly activating, giving rate ratios of 2.9 and 3.7, respectively.

The damping effect of the isopropyl and *t*-butyl groups is in all likelihood steric in origin. The activating effect of the nitro and cyano groups, on the other hand, is explicable in terms of the possibility for delocalization of the unpaired electron in the radical intermediate (1) for *para* substitution in nitrobenzene, which is also operative in *ortho* substitution but not in *meta* substitution, and hence facilitation of attack at the *ortho* and *para* positions, and as a result, activation of the nucleus as a whole.

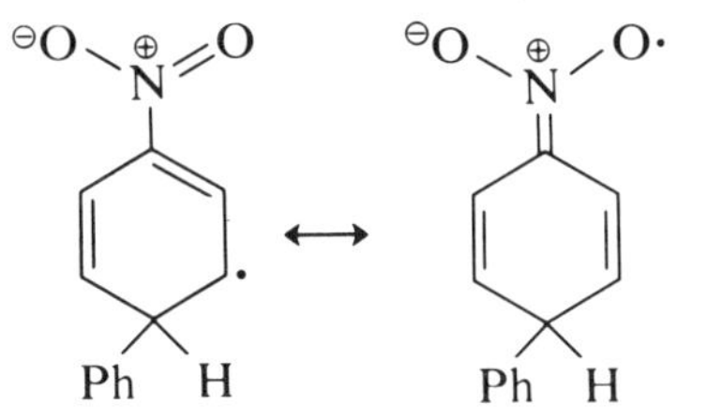

$^{\ominus}$O–$\overset{\oplus}{N}$=O ... Ph H $\longleftrightarrow$ $^{\ominus}$O–$\overset{\oplus}{N}$–O· ... Ph H

(1)

That the reaction is sensitive to polar effects was first indicated by the isomer ratios obtained in the arylation of benzotrichloride with the *p*-chloro- and *p*-nitrophenyl radicals, with which attack at the *meta* position was observed to be substantially increased, relative to that obtained with the unsubstituted phenyl radical. A large number of substituted phenyl radicals, derived from the appropriate aroyl peroxides, have since been studied, with nitrobenzene as the substrate, the relative rates and partial rate factors being determined[2] in each case. The results, shown in Table 13.2, bespeak the participation of both polar and steric effects. (It has been pointed out that nitrobenzene is not a suitable standard substrate for the study of partial rate factors in homolytic aromatic substitution in the presence of substances carrying transferable hydrogen atoms.[28]) Thus the reduced rate ratio, as well as the small but distinct increase in *meta* substitution, leave little doubt as the assertion of the electrophilic character of the nitro- and halogenophenyl radicals. The reduced activity at the *ortho* positions and

**Table 13.2** Relative Rates and Partial Rate Factors for the Arylation of Nitrobenzene at 80 °C*

| Arylating Radical | Relative Rate | Partial Rate Factors $F_o$ | $F_m$ | $F_p$ |
|---|---|---|---|---|
| $o$-$NO_2$—$C_6H_4\cdot$ | 0.26 | 0.42 | 0.14 | 0.42 |
| $m$-$NO_2$—$C_6H_4\cdot$ | 0.43 | 0.68 | 0.73 | 0.75 |
| $p$-$NO_2$—$C_6H_4\cdot$ | 0.94 | 1.6 | 0.43 | 1.6 |
| $o$-Cl—$C_6H_4\cdot$ | 0.82 | 0.88 | 0.60 | 2.0 |
| $m$-Cl—$C_6H_4\cdot$ | 1.3 | 2.2 | 0.58 | 2.2 |
| $p$-Cl—$C_6H_4\cdot$ | 1.5 | 2.7 | 0.63 | 2.5 |
| $o$-Br—$C_6H_4\cdot$ | 0.79 | 0.83 | 0.59 | 1.9 |
| $p$-Br—$C_6H_4\cdot$ | 1.8 | 3.1 | 0.70 | 2.9 |
| $C_6H_5\cdot$ | 2.9 | 5.5 | 0.86 | 4.9 |

* Data from D. H. Hey, *Advances in Free-Radical Chemistry*, 1967, **2**, G. H. Williams, Ed., Logos, London.

the resultant overall dimunition of the relative rate, when the reacting radical is an *ortho* substituted phenyl radical, can obviously be attributed to steric hindrance.

## 13.3 Mechanism of aromatic arylation

Before more precise information became available arylation of benzene was considered to proceed by the discrete steps indicated below:

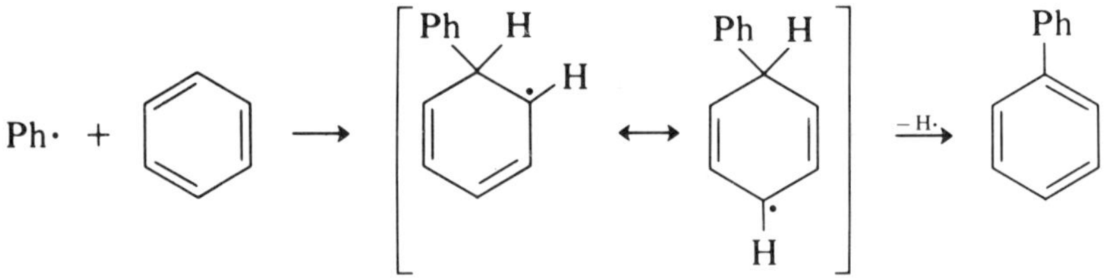

$$\text{Ph}\cdot + \text{C}_6\text{H}_6 \longrightarrow [\text{Ph}-\text{C}_6\text{H}_6]\cdot \ (2) \xrightarrow{-\text{H}\cdot} \text{Ph}-\text{C}_6\text{H}_5$$

(2)

Of these the first is one of addition to give the phenylcyclohexadienyl radical (2), referred to subsequently as $[\text{Ph}-\text{C}_6\text{H}_6]\cdot$, and this is followed by elimination of a hydrogen atom, through abstraction by another radical, to give the biaryl. More recent, comprehensive work on the decomposition of benzoyl peroxide in benzene has now established the following scheme of reactions:

(a) $(PhCO)_2O_2 \longrightarrow 2PhCOO\cdot \longrightarrow 2Ph\cdot + 2CO_2$

(b) $Ph\cdot + PhH \longrightarrow [Ph{-}C_6H_6]\cdot$

(c) $[Ph{-}C_6H_6]\cdot \xrightarrow{-H\cdot} Ph{-}Ph$

(d) $2[Ph{-}C_6H_6]\cdot \longrightarrow Ph{-}C_6H_6{-}C_6H_6{-}Ph$

(e) $2[Ph{-}C_6H_6]\cdot \longrightarrow PhC_6H_7 + Ph{-}Ph$

(f) $(Ph{-}C_6H_6{-})_2 \xrightarrow{-4H\cdot} Ph{-}C_6H_4{-}C_6H_4{-}Ph$

(g) $Ph{-}C_6H_7 \xrightarrow{-2H\cdot} Ph{-}Ph$

Of these (b) represents the addition step to form the phenylcyclohexadienyl radical (2), and (c) its subsequent dehydrogenation to the diaryl, as mentioned above. Among the additional steps established (d) and (e) represent the dimerization and disproportionation, respectively, of the radical (2) to give hydroaromatic compounds which in turn are dehydrogenated according to (f) and (g), to yield biphenyl and quaterphenyls. These reactions are illustrated below for one of the resonance hybrids of the radical (2):

$$2\ (\text{Ph},\text{H})C_6H_4\dot{}{-}H \xrightarrow{(d)} \text{Ph(H)}C_6H_4\text{(H)}{-}\text{(H)}C_6H_4\text{(Ph)(H)} \xrightarrow[-4H\cdot]{(f)} Ph{-}C_6H_4{-}C_6H_4{-}Ph$$

$$2\ (\text{Ph},\text{H})C_6H_4\dot{}{-}H \xrightarrow{(e)} \text{Ph(H)}C_6H_4\text{(H)(H)} + Ph{-}Ph$$

$$\text{Ph(H)}C_6H_4\text{(H)(H)} \xrightarrow[-2H\cdot]{(g)} Ph{-}Ph$$

The more important pieces of evidence supporting the postulation of the above scheme are listed below:

(i) In the decomposition of benzoyl peroxide in benzene, quaterphenyls were obtained in a greater yield than terphenyl. This result indicates that the terphenyl and quaterphenyl were not formed by successive phenylation.[12]

(ii) The quaterphenyl obtained from the reaction of symmetrically disubstituted benzoyl peroxide with benzene revealed substituents in only two of the aromatic nuclei.[13]

(iii) When decomposition of a very dilute solution of benzoyl peroxide in benzene was carried out in the complete absence of oxygen, hydroaromatic products could be isolated. These were dihydrobiphenyl (or a mixture of its isomers) and an isomer of tetrahydro-*p*-quaterphenyl.[14] When oxygen was not rigidly excluded, these hydroaromatic compounds could not be isolated, but were apparently oxidized to the corresponding bi- and polyphenyls. This function of oxygen could be taken over by *o*-chloranil, in the presence of which the hydroaromatic substances formed are similarly dehydrogenated.[15]

(iv) In the decomposition of benzoyl peroxide in benzene-1-$^{14}C$, while a little *p*-terphenyl labelled in one nucleus only was found, a much larger yield was obtained of 4,4′-quaterphenyl labelled in two nuclei.[16]

The conclusions derived from product studies of the decomposition of benzoyl peroxide in benzene have, in the main, received corroboration from findings in kinetic studies. These, in addition, have yielded more precise information on the nature of the termination steps (between like and unlike radicals) as well as revealed the occurrence of two modes of induced decomposition, represented by (h) and (i) below:

(h) $Ph\cdot + (PhCO)_2O_2 \rightarrow PhCO_2Ph + PhCO_2\cdot$

(i) $[Ph{-}C_6H_6]\cdot + (PhCO)_2O_2 \rightarrow Ph_2 + PhCO_2H + PhCO_2\cdot$

Thus the kinetic equation[17] obtained for the primary homolysis process of benzoyl peroxide (represented by P) into two benzoyloxy radicals, *viz.*,

$$-d[\mathrm{P}]/dt = k_1[\mathrm{P}] + k_2[\mathrm{P}]^{\frac{3}{2}}$$

contains the term $k_2[\mathrm{P}]^{\frac{3}{2}}$, attributable to induced decomposition. This gives rise to variations in the observed rate of decomposition from one solvent to another, although in simple aromatic solvents these variations are minor. Of the two modes of induced decomposition (h) and (i), recent investigations have shown the latter, the oxidation of the phenylcyclohexadienyl radicals to diaryls by molecular benzoyl peroxide, to be the one predominating. Such a process would give 1.5–order kinetics provided chains are terminated by reaction between like radicals (by dimerization or disproportionation), but not by reaction between unlike radicals.

DeTar[18] has recently made a study of the elementary steps involved in the reaction of benzoyl peroxide with the benzene nucleus using computer techniques. From cumulative information on product distribution from the thermal decomposition of the peroxide in a number of aromatic solvents, the known absolute rate constants of representative radical reactions, and the limited data available on the kinetics of reactions of the peroxide, he has been able to assign rate constants to most of the elementary steps in the mechanism generally accepted.

## 13.4 Aromatic alkylation[19]

Alkylation is usually more complex than arylation and the yields generally poor, products containing vulnerable benzylic hydrogen atoms being particularly susceptible to side reactions. In addition, there is a greater tendency for the alkyl cyclohexadienyl radical intermediate to undergo disproportionation and/or dimerization, as shown.

$$R\cdot + PhH \longrightarrow [\text{R, H-substituted cyclohexadienyl radical} \longleftrightarrow \text{R, H-substituted cyclohexadienyl radical}] \longrightarrow PhR; \quad R{-}C_6H_6{-}C_6H_6{-}R$$

*ortho* and *para* dimers

For generation of the alkyl radicals various methods have been found applicable, as outlined below (through thermolysis except where otherwise indicated):

1. $(RCO)_2O_2 \rightarrow 2RCOO\cdot \rightarrow 2R\cdot + 2CO_2$
2. $RCOO^- \xrightarrow{-e} RCOO\cdot \rightarrow R\cdot + CO_2$
3. $RHgI \xrightarrow{h\nu} R + HgI$
4. $Pb(OCOR)_4 \rightarrow Pb(OCOR)_2 + 2RCOO\cdot \rightarrow 2R\cdot + 2CO_2$
5. $PhI(OCOR)_2 \rightarrow PhI + 2RCOO\cdot \rightarrow 2R\cdot + 2CO_2$
6. $CH_3CON(NO)CH_2Ph \rightarrow PhCH_2\cdot$

Quantitative study of homolytic alkylation has been undertaken. The isomer ratios obtained for the methyl radical, as given in Table 13.3, reveal the same overall pattern of substitution as in phenylation. The nitro and cyano groups, which conjugate extensively with the aromatic ring, in fact give rise to greater selectivity at the *ortho* and *para* positions in the methylation than in the phenylation reaction. This conjugation effect would be expected to enhance reactivity, as has indeed been substantiated by the high methyl affinity of benzonitrile (for nitrobenzene, the relative rate value is not yet available). It is also noteworthy that the proportion of *ortho* isomer formed is significantly higher in methylation than in phenylation reactions, a result suggesting relative freedom from steric hindrance.

In a study by Waters, *et al.*,[20] of the methylation and phenylation of a series of substituted benzenes, a comparison of the yields of *meta* and *para* isomers formed, referred to as the *meta-para* ratio, has given indications of

**Table 13.3** Isomer Ratios in Homolytic Methylation*

| Aromatic Compound Methylated | Product (%) | | | Temperature (°C) |
|---|---|---|---|---|
| | *ortho* | *meta* | *para* | |
| Chlorobenzene | 64 | 25 | 11 | 130 |
| Bromobenzene | 67.5 | 23 | 9.5 | 145 |
| Anisole | 74 | 15 | 11 | 140 |
| Nitrobenzene | 65.5 | 6 | 28.5 | 143 |
| Benzonitrile | 48 | 9 | 43 | 133 |
| Toluene | 56.5 | 26.5 | 17 | 110 |

* Data from G. H. Williams, *Homolytic Aromatic Substitution,* Pergamon Press, London, 1960.

the relative nucleophilicity of the methyl and phenyl radicals. (Results on substitution at the *ortho* position have not been used because of likely steric effects). The data listed in Table 13.4 reveal that, for substituent groups which are *ortho-para* directing in heterolytic substitution reactions, the said ratio is higher for methylation than it is for phenylation. Since such substituents are known, through inductive and resonance effects, to bring about a high electron density at the *para* position relative to that at the *meta* position, this result indicates that the methyl radical is more nucleophilic than the phenyl radical. Conversely, for substituents of the opposite polar character the reverse situation is observed, the *meta-para* ratio here being decidedly lower for methylation than for phenylation, once more pointing to stronger nucleophilic character in the methyl radical. This work has been

**Table 13.4** *Meta-para* Ratios for Homolytic Substitution of PhX*

| X | Phenylation | Methylation |
|---|---|---|
| $CH_3$ | 1.32 | 1.56 |
| Cl | 1.71 | 2.27 |
| Br | 1.91 | 2.42 |
| OMe | 1.20 | 1.36 |
| $NO_2$ | 0.32 | 0.21 |
| CN | 0.81 | 0.21 |

* Data from G. H. Williams, *Homolytic Aromatic Substitution,* Pergamon Press, London, 1960.

extended to other alkyl radicals such as ethyl and *n*-butyl which were found to be more nucleophilic then aryl radicals.[29]

## 13.5 Intramolecular aromatic substitution

Interest in radical cyclization reactions has steadily increased during the past decade,[21,22] the most common examples being addition reactions to olefinic bonds and to aromatic nuclei. Some of these have found application in the construction of complex molecules, advantage being often taken of the facilitation of reaction by liberating the radical centre in close proximity to the desired position of attack.

The best known among intramolecular aromatic arylation processes is probably the Pschorr Synthesis, developed as early as 1896, and reactions related to it, on which a good deal of work has been done. The reaction has been applied to the synthesis of a large number of aromatic systems, and may be represented as follows:

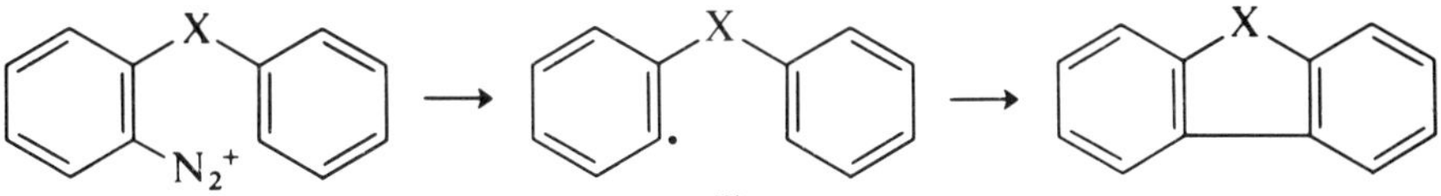

X $N_2^+$ → X · → X

(3)

X = $CH_2$, O, CH=CH (*cis*), CO, $CH_2CH_2$, NR, S or $SO_2$

The diazonium ion is generally decomposed using copper powder or copper salts, to liberate the substituted phenyl radical (3) which then attacks the proximate aromatic nucleus. Hey and his co-workers[23] have studied the effect of the nuclear substituents R on the nucleus undergoing attack, in a variety of systems of which (4) is typical, but have found that such substituents hardly affect the yields of the cyclized product. This is so even with the nitro group, which is known to activate the aromatic ring towards radical arylation.

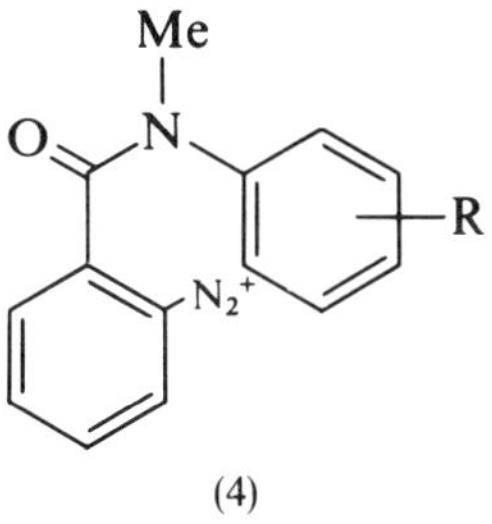

Me N O R $N_2^+$

(4)

On the other hand, copper catalysed decomposition of the diazonium salt from the amine (5) gives a 90% yield of the isomeric fluorenones (6) and (7), possibly due to the reinforcing activating influences of the nitro and the carbonyl groups.

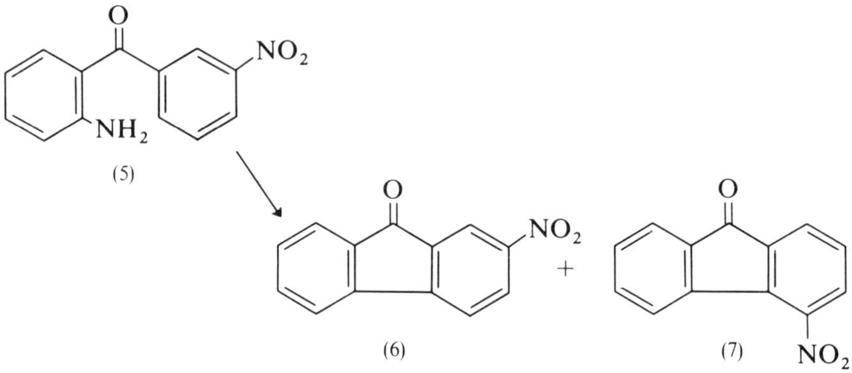

O
$NO_2$
$NH_2$
(5)
O
$NO_2$
+
O
(6)
(7) $NO_2$

Other methods of generating the arylating radical may also be used. For example, electrolysis of the sodium salt of *o*-benzoylbenzoic acid, or persulphate oxidation of the acid itself, gives, via the *o*-benzoylphenyl radical, a mixture of fluorenone, fluoren-9-ol, and benzophenone.[24] Kharasch's cobaltous chloride method, when applied[25] to the aryl bromide (8), gives the phenanthrene (9).

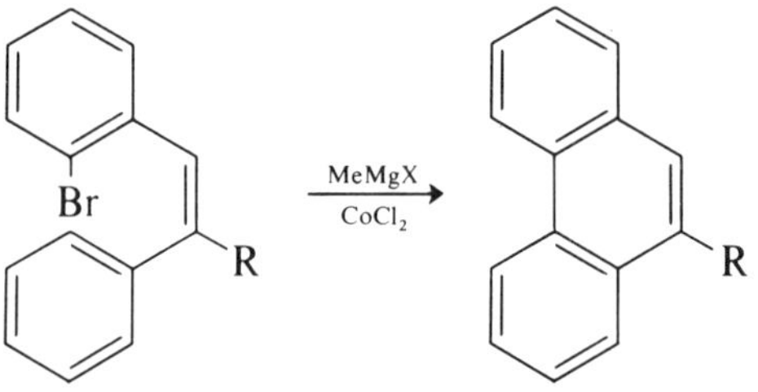

Br
R
MeMgX
$CoCl_2$
R
(8)
(9)

Very few instances of intramolecular alkylation have so far been encountered. The 4-phenylbutyl radical, when generated in a solvent which is a relatively poor transfer agent, has been found to undergo cyclization. The radical was generated in a variety of ways: by decarbonylation of the appropriate aldehyde, by decomposition of the requisite acyl peroxide, or by electrolysis of a salt of 5-phenylpentanoic acid or treatment of the acid with lead tetra-acetate. A more recent example is the formation of anthraquinone from the radical (10), derived from *o*-benzoyltoluene through hydrogen abstraction by *t*-butoxy radicals.[26]

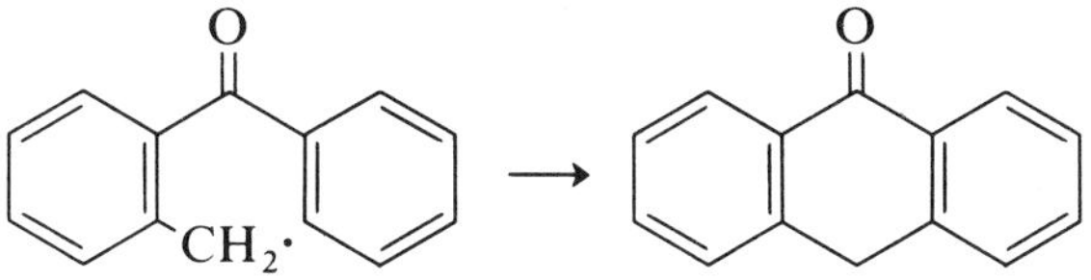

(10)

Among the few reported occurrences of intramolecular acylation may be quoted the cyclization of $\omega$-aryl propionyl and -butyroyl radicals to indanones and tetralones (illustrated below for the former):

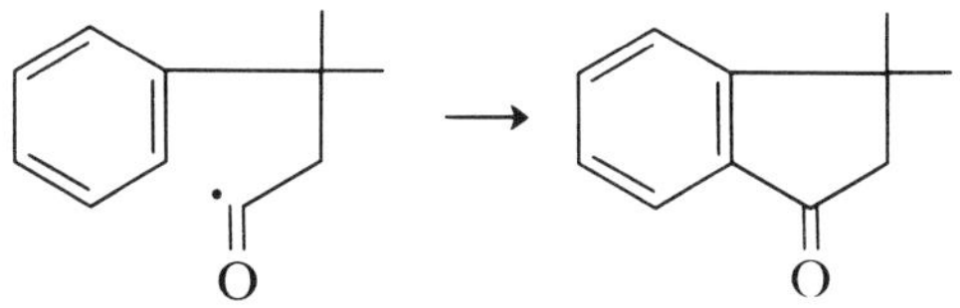

A similar cyclization to a fluorenone has also been recorded:[26]

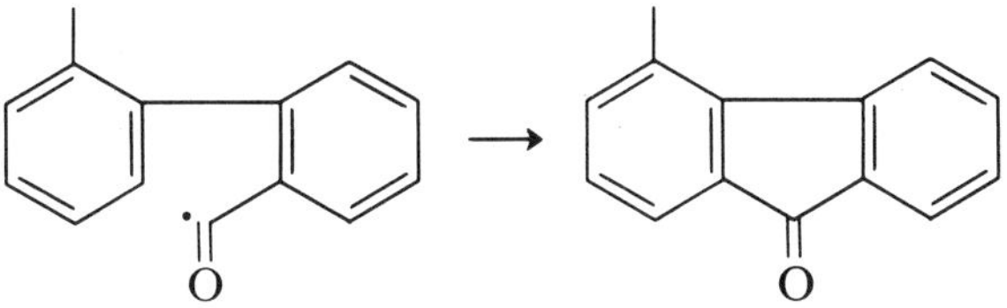

## References

1 W. S. M. GRIEVE and D. H. HEY, *J. chem. Soc.*, 1934, 1797.
2 D. H. HEY, *Adv. Free-Radical Chem.*, 1967, **2**, G. H. WILLIAMS, Ed., Logos, London.
3 J. I. G. CADOGAN, *J. chem. Soc.*, 1962, 4257; HWANG SHU, *Acta chim. sin.*, 1959, **25**, 171.
4 J. I. G. CADOGAN, *Acc. Chem. Res.*, 1971, **4**, 186.
5 H. GELISSEN and P. H. HERMANS, *Chem. Ber.*, 1925, **58**, 285.
6 D. H. HEY, *J. chem. Soc.*, 1934, 1966; H. WIELAND, S. SCHAPIRO, and H. METZGER, *Justus Liebigs Annln. Chem.*, 1934, **513**, 93.
7 M. EBERHARDT and E. L. ELIEL, *J. org. Chem.*, 1962, **27**, 2289.
8 G. R. CHALFONT, D. H. HEY, K. S. Y. LIANG, and M. J. PERKINS, *J. chem. Soc. (B)*, 1971, 233.
9 W. A. WATERS, *Vistas in Free-Radical Chemistry*, Pergamon, London, 1959, p. 3.
10 D. I. DAVIES, M. TIECCO, and D. H. HEY, *J. chem. Soc.*, 1965, 7062.
11 G. M. BADGER and C. P. WHITTLE, *Aust. J. Chem.*, 1963, **16**, 440.
12 B. M. LYNCH and K. H. PAUSACKER, *Aust. J. Chem.*, 1957, **10**, 40.
13 K. H. PAUSACKER, *Aust. J. Chem.*, 1957, **10**, 49.
14 D. F. DETAR and R. A. J. LONG, *J. Am. chem. Soc.*, 1958, **80**, 4742.
15 D. H. HEY, M. J. PERKINS, and G. H. WILLIAMS, *J. chem. Soc.*, 1963, 5604.
16 G. A. RAZUVAEV, B. G. ZATEEV, and G. G. PETUKHOV, *Dokl. Akad. Nauk SSSR*, 1960, **130**, 336.
17 K. NOZAKI and P. D. BARTLETT, *J. Am. chem. Soc.*, 1946, **68**, 1686; K. NOZAKI and P. D. BARTLETT, *J. Am. chem. Soc.*, 1947, **69**, 2299.
18 D. F. DETAR, *J. Am. chem. Soc.*, 1967, **89**, 4058.

19 G. H. WILLIAMS, *Homolytic Aromatic Substitution*, Pergamon Press, London, 1960.
20 B. R. COWLEY, R. O. C. NORMAN, and W. A. WATERS, *J. chem. Soc.*, 1959, 1799.
21 MARC JULIA, Free Radical Cyclisations, in *International Symposium on Free Radicals in Solution*, Butterworth, London, 1967, p. 167.
22 R. A. ABRAMOVITCH, *Adv. Free-Radical Chem.*, 1967, **2**, 87, G. H. WILLIAMS, Ed., Logos, London.
23 R. A. HEACOCK and D. H. HEY, *J. chem. Soc.*, 1952, 1508, 4059; 1953, 3.
24 P. J. BUNYAN and D. H. HEY, *Proc. chem. Soc.*, 1959, 366.
25 M. TIECCO, *Chem. Commun.*, 1965, 555.
26 R. L. HUANG and H. H. LEE, *J. chem. Soc.* (*C*), 1966, 929.
27 D. I. DAVIES, D. H. HEY, and B. SUMMERS, *J. chem. Soc.* (*C*), 1971, 2681.
28 H. OHTA and K. TOKUMARU, *Bull. Chem. Soc. Japan*, 1971, **44**, 3218.
29 G. E. CORBETT and G. H. WILLIAMS, *J. chem. Soc.* (*B*), 1966, 877.

# 14
# Radical Fragmentation

Fragmentation processes, sometimes referred to as $\beta$ elimination, may be looked upon as the reverse of addition reactions. Typical examples are given below:

$$R_3C{-}O\cdot \rightarrow R_2C{=}O + R\cdot$$

$$R_2\dot{C}{-}OR \rightarrow R_2C{=}O + R\cdot$$

$$R{-}C({=}O){-}O\cdot \rightarrow CO_2 + R\cdot$$

$$>C(X){-}\dot{C}< \rightarrow >C{=}C< + X\cdot$$

$$(X = \text{Br, I, or SR})$$

These processes are usually endothermic and are favoured by an increase in temperature. The major driving force is derived from the resulting increase in entropy since the formation of two species from one brings about additional translational degrees of freedom. Fragmentation of a cyclic radical results also in an entropy increase, in this case due to enhancement in rotational degrees of freedom.

### 14.1 Fragmentation of alkoxy radicals

Most of the information now available on alkoxy radicals has been due to Walling.[1] $\beta$ Elimination in these entities is much more facile than in alkyl radicals, and is due to formation of the $\pi$-bond in the resulting carbonyl group; for example, the enthalpy for fragmentation of neopentyl ($\Delta H \sim 20$ kcal mole$^{-1}$) is more endothermic than that for the *t*-butoxy radical ($\Delta H \sim 5$ kcal mole$^{-1}$).

The *t*-butoxy radical (from di-*t*-butyl peroxide) has been extensively studied. In a donor solvent it readily abstracts a hydrogen atom to give *t*-butanol, otherwise it undergoes fragmentation to acetone and a methyl radical. This latter process is favoured at higher temperatures: thus at 125 °C *t*-butoxy radicals in various solvents are known to give *t*-butanol and acetone in a ratio of 4.1 while at 145 °C in the same solvents this ratio is decreased to 1.6. Aromatic solvents have also been found to lower this ratio and this has been attributed to complexing of the radicals with the solvent. Solvent complexing has the effect not only of inhibiting bimolecular hydrogen abstraction but also of lowering the energy of the transition state involved in fragmentation, hence decreasing the *t*-butanol–acetone ratio.

The effect of structure on the ease of fragmentation of the *t*-alkoxy radicals (1), where R is $Bu^t$, $Pr^i$, Et, $PhCH_2$, Ph, $ClCH_2$, and Me, has been studied at 40 °C by comparing, for each radical, the extents to which the two processes of fragmentation and of hydrogen abstraction occur.[1] The facility with which the former process takes place has thus been found to follow approximately the order given above. This is reminiscent of the α-alkoxybenzyl radicals PhCHOR (from the benzyl ethers $PhCH_2OR$) studied earlier,[2] the tendencies of which to break down to benzaldehyde and R· also follow much the same order (see Section 3.4). The particular ease with which the *t*-butyl radical is eliminated suggests that not only is the stability of the newly formed radical important, but its size is also a contributing factor. The relatively less facile elimination of free benzyl, on the other hand, seems to indicate that bond

$$\mathrm{R{-}C(Me)_2{-}O\cdot} \longrightarrow \mathrm{Me_2C{=}O + R\cdot}$$

$$\mathrm{R{-}C(Me)_2{-}O\cdot} \xrightarrow{\text{cyclohexane}} \mathrm{RCMe_2OH + C_6H_{11}\cdot}$$

(1)

$$\mathrm{R_1R_2R_3C{-}O\cdot} \longrightarrow \mathrm{R_1\cdot + R_2COR_3}$$

$$\mathrm{R_1R_2R_3C{-}O\cdot} \longrightarrow \mathrm{R_2\cdot + R_1COR_3}$$

$$\mathrm{R_1R_2R_3C{-}O\cdot} \longrightarrow \mathrm{R_3\cdot + R_1COR_2}$$

(2)

$$\left[\overset{\delta^+}{\mathrm{R_2}}\cdots\underset{\mathrm{R_3}}{\overset{\mathrm{R_1}}{\mathrm{C}}}\cdots\overset{\delta^-}{\mathrm{O}}\right]^{\cdot}$$

(3)

breaking occurs early in the transition state and, consequently, stabilization by benzylic resonance is not called into play.

In separate studies on the cleavage of the alkoxy radicals (2) Kochi[3a] and Greene *et al.*[3b] have also found that the course of breakdown is determined by the relative stability of the radical to be formed. Thus the fragmentation of *t*-amyloxy and 2-phenyl-2-butoxy radicals gives mainly ethyl radicals. Fragmentation of these alkoxy radicals (2) is much more selective than that of radicals derived from acetals (discussed below). It has been suggested that the transition state is probably strongly polarized, as in (3).

Radicals generated by hydrogen abstraction of dibenzyl phenylamines,[4a] benzaldehyde dimethylacetal,[4b] and triethyl orthoformate [4c] have also been found to undergo extensive fragmentation, as shown.

$$\mathrm{Ph\dot{C}HN(Ph){-}CH_2Ph \rightarrow PhCH{=}NPh + PhCH_2\cdot}$$

$$\mathrm{Ph\dot{C}(OMe)_2 \rightarrow PhCO_2Me + Me\cdot}$$

$$\mathrm{EtO\dot{C}(OEt)_2 \rightarrow EtO{-}C({=}O){-}OEt + Et\cdot}$$

Huyser and Wang[5] studied the fragmentation of the radicals (4) derived from mixed acetals and found that the major factor governing the course of fragmentation is the stability of the radical formed. Resonance stabilized radicals such as free benzyl and allyl are most readily expelled. Tertiary radicals are also eliminated with ease, followed by secondary and primary radicals. The cyclo-octyl group similarly breaks off readily probably due to the consequent relief of conformational strain present in a medium sized ring.

$$\underset{(4)}{\mathrm{Me{-}\dot{C}(OR)(OBu^n)}} \rightarrow \mathrm{Me{-}C({=}O){-}OBu^n + R\cdot} \quad \text{or} \quad \mathrm{Me{-}C({=}O){-}OR + Bu^n\cdot}$$

## 14.2 Fragmentation in cyclic radicals

Breakdown in cyclic radicals usually results in rearranged products. The addition of the trichloromethyl radical to $\beta$-pinene, for instance, leads to ring opening, as shown.[6] Relief of ring strain here complements the entropy gain in such reactions: Ring opening is observed in the reaction of cyclobutyl methyl iodide initiated by benzoyl peroxide but not in the reaction initiated by triphenyltin iodide.[7] In the latter case the intermediate cyclobutyl methyl radical is apparently trapped by the organotin reagent before it can ring open to the 4-pentenyl radical.

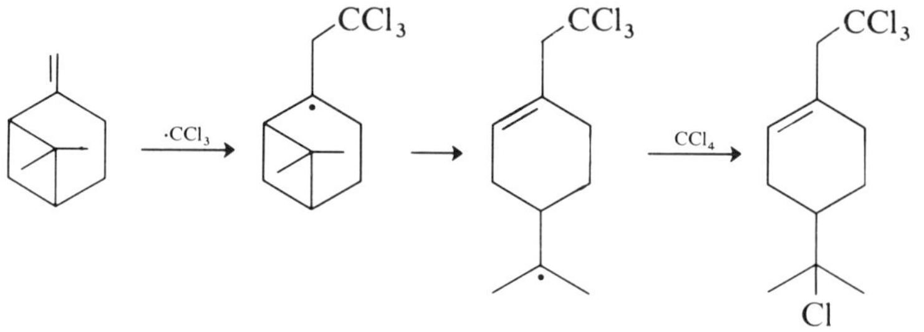

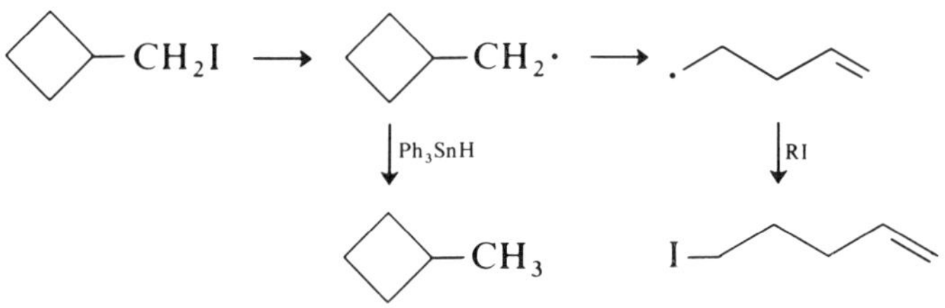

$CCl_3$ $CCl_3$ $CCl_3$

$\cdot CCl_3$ $CCl_4$

Cl

$CH_2I$ → $CH_2\cdot$ →

$Ph_3SnH$ RI

$CH_3$ I

$\beta$-Scission leading to the ring opening of three-membered rings is also known. This may be illustrated by the following example of the opening of an epoxide ring. Fragmentation of this type is similar to the cyclopropyl-carbinylhomoallylic radical rearrangement which is treated in more detail in Chapter 16.

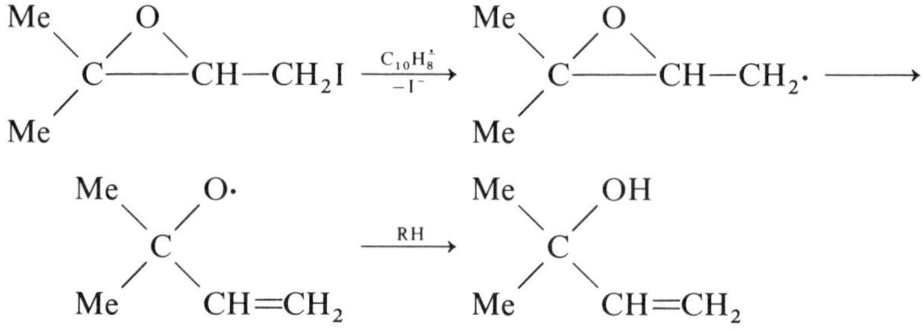

$$\mathrm{Me_2C\overset{O}{—}CH{-}CH_2I \xrightarrow[-I^-]{C_{10}H_8^{\cdot-}} Me_2C\overset{O}{—}CH{-}CH_2\cdot \longrightarrow}$$

$$\mathrm{Me_2C(O\cdot)CH{=}CH_2 \xrightarrow{RH} Me_2C(OH)CH{=}CH_2}$$

The cyclic acetal, 2-phenyl-1,3-dioxalane, undergoes peroxide induced fragmentation by a chain process to yield ethyl benzoate.[8]

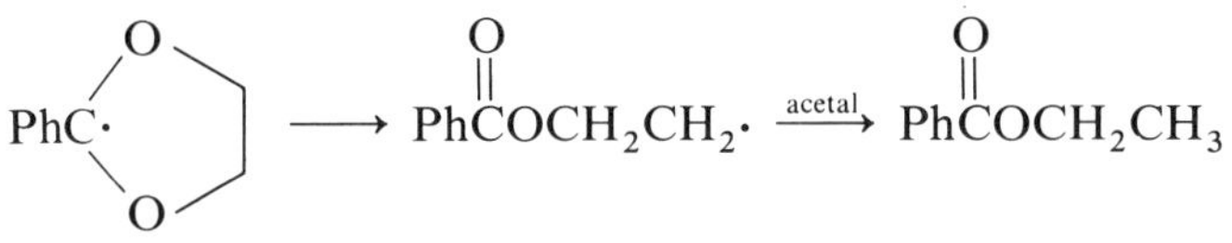

$$\text{PhC}\cdot\text{(OCH}_2\text{CH}_2\text{O)} \longrightarrow \text{Ph}\overset{\text{O}}{\overset{\|}{\text{C}}}\text{OCH}_2\text{CH}_2\cdot \xrightarrow{\text{acetal}} \text{Ph}\overset{\text{O}}{\overset{\|}{\text{C}}}\text{OCH}_2\text{CH}_3$$

An unusual rearrangement has been observed in the thermal decomposition of the azo compounds (5) and (6), as illustrated below (Berson *et al.*):[9]

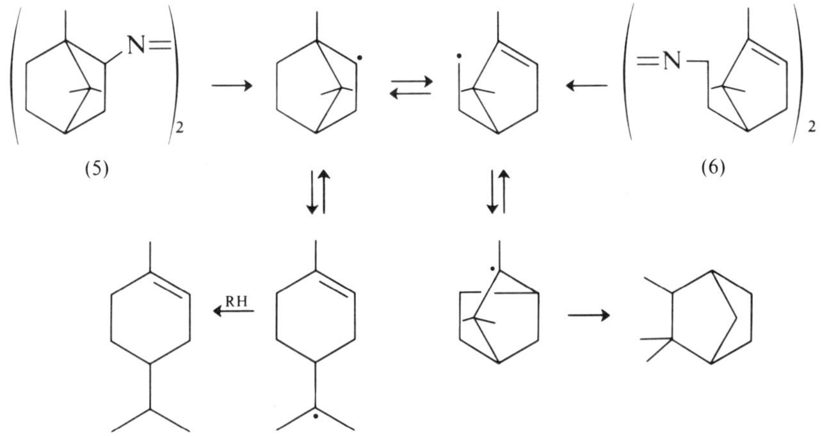

(5) (6) RH

## 14.3 Fragmentation of acyl and acyloxy radicals

The decarboxylation of acyloxy radicals and the decarbonylation of acyl radicals are well known routes to the radical R·, as shown:

$$\text{RCO}_2\cdot \rightarrow \text{R}\cdot + \text{CO}_2$$

$$\text{R}\dot{\text{C}}\text{O} \rightarrow \text{R}\cdot + \text{CO}$$

Acyloxy radicals are readily generated from diacyl peroxides, which in turn can be prepared from the corresponding carboxylic acids. Alternatively carboxylate anions can be oxidized electrolytically to yield the acyloxy radicals. Acyl radicals are usually obtained from the aldehydes by abstraction of the labile hydrogen atoms, or through irradiation. The fragmentation of acyl radicals constitutes a convenient route to alkyl radicals while that of acyloxy radicals can provide aryl radicals as well.

## 14.4 Fragmentation of hydroperoxyalkyl radicals

Fragmentation of intermediate radicals from the oxidation of hydrocarbons occurs frequently.[10] In fact the production of alkenes by oxidation of saturated hydrocarbons depends on such a process. For illustration, the gaseous oxidation of isobutane may be written:

$$Me_2CHCH_3 \nearrow \quad Me_2C(CH_2{-}H)(O{-}\dot{O}) \rightarrow Me_2C(\dot{C}H_2){-}O{-}O{-}H \searrow \quad Me_2C{=}CH_2 + HO_2\cdot$$

$$Me_2CHCH_3 \searrow \quad Me_2C(H){-}CH_2{-}O{-}O\cdot \rightarrow Me_2\dot{C}CH_2OOH \nearrow \quad Me_2C{=}CH_2 + HO_2\cdot$$

More complicated cases of fragmentation of hydroperoxyalkyl radicals are known. Of these a few examples are given below.

$$HO{-}OCH_2{-}CH(Me){-}\dot{C}HEt \rightarrow HO\cdot + CH_2O + MeCH{=}CHEt$$

$$Me_2C(CH_2{-}\dot{C}HMe)(O{-}OH) \rightarrow Me_2C{=}O + CH_2{=}CHMe + \cdot OH$$

$$Me_2C(O{-}OH){-}CH_2{-}CH_2{-}CH_2\cdot \rightarrow Me_2C\underset{O}{\diagdown\diagup}CH_2 + CH_2{=}CH_2 + \cdot OH$$

### References


1 C. WALLING, *Free Radicals in Solution*, Butterworth, London, 1967, p. 69.

2 R. L. HUANG and S. S. SI-HOE, *Vistas in Free Radical Chemistry*, Ed. W. A. WATERS, Pergamon, 1959, p. 242; *Proc. chem. Soc.*, 1957, 354.

3*a* J. K. KOCHI, *J. Am. chem. Soc.*, 1963, **85**, 1593;

*b* F. D. GREENE, M. L. SAVITZ, F. D. OSTERHOLTZ, H. H. LAU, W. N. SMITH, and P. M. ZANET, *J. org. Chem.*, 1963, **28**, 55.

4*a* R. L. HUANG, *J. chem. Soc.*, 1959, 1816;

*b* R. L. HUANG and K. H. LEE, *J. chem. Soc.*, 1964, 5957;

*c* E. S. HUYSER and D. T. WANG, *J. org. Chem.*, 1962, **27**, 4696.

5 E. S. HUYSER and D. T. WANG, *J. org. Chem.*, 1964, **29**, 2720.

6*a* D. M. Oldroyd, G. S. Fisher, and L. A. Goldblatt, *J. Am. chem. Soc.*, 1950, **72**, 2407;
*b* G. Du Pont, R. Dulon, and G. Clement, *Bull. Soc. chim. Fr.*, 1950, 1056, 1115.
7 L. Kaplan, *J. org. Chem.*, 1968, **33**, 2531.
8 E. S. Huyser and Z. Garcia, *J. org. Chem.*, 1962, **27**, 2716.
9 J. A. Berson, C. J. Olsen, and J. S. Walia, *J. Am. chem. Soc.*, 1962, **84**, 3337.
10 A. Fish, *Organic Peroxides*, Ed. D. Swern, Wiley, 1970, Vol. 1, Chapter 3.

# 15
# Dimerization of Radicals

Dimerization is a radical destroying process and constitutes one of the termination pathways in chain reactions. Since bond making is exothermic the molecule so formed must, to achieve stability, dissipate the energy generated either by spreading it throughout its vibrational modes, or through collision with a 'third body' such as the wall of the reaction vessel or another molecule present in the reaction system. Consider the combination of two radicals X· and Y· to form an initially excited species XY*, in the presence

$$X\cdot + Y\cdot \underset{k_2}{\overset{k_1}{\rightleftharpoons}} XY^*$$

$$XY^* + M \xrightarrow{k_3} XY + M^*$$

of a 'third body' M. When XY* reaches the steady state the rate of formation of the stable molecule XY will be

$$\frac{d[XY]}{dt} = \frac{k_1k_3[X\cdot][Y\cdot][M]}{k_2 + k_3[M]}$$

There are two cases to be considered. At low concentrations of M, $k_2 \gg k_3[M]$, and the rate of formation of XY equals $(k_1k_3/k_2)[X\cdot][Y\cdot][M]$. At high concentrations of M, where $k_2 \ll k_3[M]$, this rate reduces to $k_1[X\cdot][Y\cdot]$ and becomes independent of the concentration of M·. Experimental work has confirmed this third body effect for methyl and ethyl radicals and for iodine atoms, whereas larger radicals always follow second order termination rate laws. An example is the combination of two ethyl radicals, which at very low pressures is third order but above 1 mm Hg becomes second order. The rate of third order combinations depends on the concentration of the third body and also on its structure. For example, the efficiency of various molecules in promoting the combination of two iodine atoms increases with the size and complexity of these molecules.

Since dimerization is a bimolecular process, its rate will depend on the square root of the rate of initiation, thus:

$$\text{Initiator} \xrightarrow{R_i} 2\text{R}\cdot$$

$$2\text{R}\cdot \xrightarrow{k_t} \text{Non-radical products}$$

At the steady state

$$R_i = 2k_t[\text{R}\cdot]^2$$

Hence

$$[\text{R}\cdot] = \left(\frac{R_i}{2k_t}\right)^{\frac{1}{2}}$$

The observed rate of the overall process is usually proportional to the concentration of radicals, and the total rate of the process therefore varies with the square root of the rate of initiation.

## 15.1 Coupling rates of radicals

Many free radicals react with each other fast enough to enable their coupling reaction to compete with diffusion from the solvent cage. Rate constants for bimolecular combination of free radicals which have been measured range from 3600 $M^{-1}\ s^{-1}$ for triphenylmethyl radicals to $3.6 \times 10^{10}\ M^{-1}\ s^{-1}$ for methyl radicals. For most of the intermediate cases direct measurement of the rate of radical-radical combination has not been made, but information derived from studies on cage effects shows that the rate constant declines only by a factor of 10 as we go from the small and relatively unstabilized methyl radical to the larger benzhydryl radical with not less than 25 kcal of resonance energy. Thus an additional benzene ring in triphenylmethyl must in one step cause the remaining millionfold ($3.6 \times 10^{10}$ to $3.6 \times 10^3$) reduction in the rate of coupling found for this radical. This obviously cannot be accounted for by resonance alone but has been attributed to steric effects. The magnitude of these effects is, in fact, such that when the triphenylmethyl radical couples with a like radical, it does so with the resonance hybrid (1) to give the isomeric dimer (2) rather than hexaphenylethane.[2]

$Ph_3C\cdot$ + H–$\dot{C}$(cyclohexadiene)=$CPh_2$ (1) ⟶ H($Ph_3C$)C(cyclohexadiene)=$CPh_2$ (2)

The rate of combination of ethyl radicals has been measured with less precision, the best value obtained being probably

$$k = 1.6 \times 10^{11}\, e^{(-2000 \pm 1000)/RT}\ \mathrm{M^{-1}\ s^{-1}}$$

When two unlike radicals combine, three routes are possible:

$$2\mathrm{X}\cdot \xrightarrow{k_{xx}} \mathrm{X_2}$$

$$2\mathrm{Y}\cdot \xrightarrow{k_{yy}} \mathrm{Y_2}$$

$$\mathrm{X}\cdot + \mathrm{Y}\cdot \xrightarrow{k_{xy}} \mathrm{XY}$$

If the relative rates depend only on statistical factors then the relative tendency $\phi$ to cross-dimerize is given by

$$\phi = \frac{k_{xy}}{(k_{xx}k_{yy})^{\frac{1}{2}}} = 2$$

The value of $\phi$ has been measured for most of the simple aliphatic radicals in the gas phase and found to be very close to 2. However, in systems in which polar effects are important, e.g., in polymerizing systems where polymeric radicals of differing charge types are involved, or in autoxidation systems where combinations between alkyl radicals R· and peroxy radicals ROO· are often favoured over coupling of like radicals, $\phi$ may depart considerably from 2.

## 15.2 Competition between coupling and disproportionation

A reaction which competes with dimerization as a termination process is that of disproportionation (see Section 3.1), which in the case of alkyl radicals leads to formation of olefins:

$$2\mathrm{R_2CH{-}\dot{C}R_2} \xrightarrow{k_d} \mathrm{R_2CH{-}CHR_2} + \mathrm{R_2C{=}CR_2}$$

(R = H or alkyl)

The ratio of the rate of disproportionation $k_d$ to that of combination $k_c$, as determined by product analysis for a number of alkyl radicals (Table 15.1) indicate that for unbranched radicals coupling is favoured over disproportionation, but that this latter process becomes more important as there is more branching in the structure of the radical.

It has been found that most polymeric radicals terminate predominantly or entirely by combination, i.e., $k_d/k_c$ is very small. However, the polymeric radical from methyl methacrylate undergoes termination by both mechanisms. The activation energy for disproportionation is about 5 kcal mole$^{-1}$

**Table 15.1** Disproportionation/Combination Ratios for Radicals[3]

| Radical Pair | Temperature (°C) | $k_d/k_c$ |
|---|---|---|
| Me· + Et· | 25–240 | 0.04 |
| Me· + Prⁿ· | 118–144 | 0.03 |
| 2Et· | 25–350 | 0.1 |
| 2Prⁿ· | 25–150 | 0.1 |
| 2Prⁱ· | 20–200 | 0.5 |
| 2Buⁱ· | 100 | 0.4 |
| 2Buˢ· | 100 | 2.3 |
| 2Buᵗ· | 100 | 4.6 |

larger than for combination, and termination by disproportionation therefore increases with increasing temperature. At 60 °C, $k_d/k_c$ is about 1.5.

## 15.3 Stereochemistry of radical coupling

When an alkyl radical XYZC· with three different substituents attached to the radical centre couples with a like radical, the course of dimerization may be as depicted in (a), which gives the *meso* dimer, or (b), which yields the racemic dimer, or both:

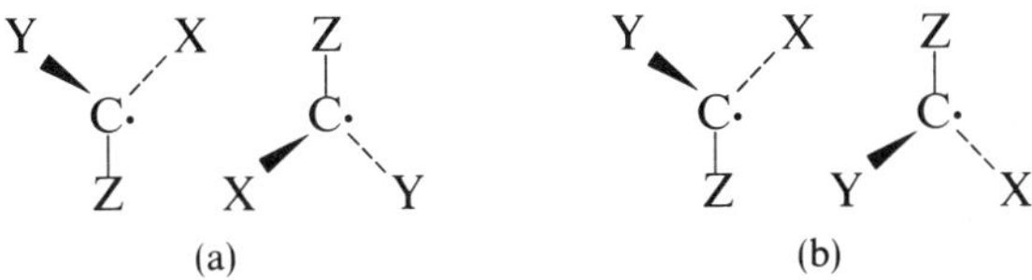

(a) (b)

Statistically, one would expect equal quantities of the diastereoisomers to be formed, as for instance is found in the dimerization[4] of the benzyl radical PhMeCH·, derived from ethylbenzene. When the substituents are polar in character or bulky in size, polar or steric effects may intervene, resulting in the preferential formation of one of the isomers. This appears to be the case with the radicals (3), (4), and (5), derived from chloroacetic acid,[4] diethyl succinate,[5]

$Cl\dot{C}HCO_2H$ (3)

$\dot{C}HCO_2Et$—$CH_2CO_2Et$ (4)

$Ph\dot{C}HCH_2CO_2Et$ (5)

and ethyl β-phenylpropionate,[6] respectively, which lead predominantly to the respective *meso* dimers. Except where near quantitative yields of diastereoisomers are realized study of the steric course of dimerization is hampered by the difficulty of estimating the exact yields of the diastereoisomers, as these usually do not differ sufficiently in their physical properties to allow quantitative separation. However, for the system Ph(OR)-CH—CH(OR)Ph, formed by the dimerization of the benzyl radicals PhĊHOR, a precise method of analysis has recently been found using n.m.r. spectroscopy, advantage being taken of the fact that the benzylic protons of the *meso* dimer appear in a higher field than those of the racemic modification.[7] For a series of radicals PhĊHOR, generated from the benzyl ethers $PhCH_2OR$ by the action of *t*-butoxy radicals (from *t*-butyl peroxide) it has been found by this method that, even with R being a relatively bulky substituent group like cyclohexyl, phenyl, *t*-butyl or 1-adamantyl, the ratio of *meso* to the racemic dimers formed was $1.0 \pm 0.05$. It is to be noted, however, that R is not directly attached to the radical centre, but through an oxygen atom.

When the coupling radicals are generated in close proximity and held together, momentarily at least, in a solvent cage one might expect a steric course of dimerization different from that of radicals generated at random, as those mentioned above. The azo compound is a system well suited for study of such a phenomenon. Thus homolysis of the azo compound XYZC—N=N—CXYZ will give two radical fragments (each denoted by R·) which might diffuse away from each other or interact within the solvent cage by coupling or by disproportionation. Each of the radicals R· may, before entering any reaction, undergo a 180° out-of-plane rotation relative to its partner and is thereby converted into its mirror image R′· (see figure below). Coupling can thus proceed with the formation of R—R and R′—R′, which are optically active, and R—R′, which is the *meso* modification of the dimer, as follows:

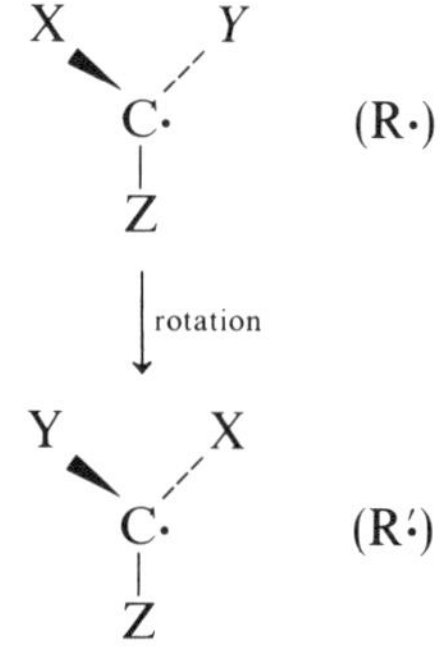

$$\begin{array}{lll}
R{-}N{=}N{-}R & & \\
\quad\downarrow & & \\
[R\cdot \quad N_2 \quad \cdot R] \rightarrow R{-}R & & \text{optically active, with retention of configuration} \\
\quad\rightleftarrows & & \\
[R'\cdot \quad N_2 \quad \cdot R] \rightarrow R'{-}R & & \textit{meso} \\
\quad\rightleftarrows & & \\
[R'\cdot \quad N_2 \quad \cdot R'] \rightarrow R'{-}R' & & \text{optically active, with inverse of configuration}
\end{array}$$

An illustration of these processes is afforded by Greene's recent work[8] on the thermal decomposition of azo bis-α-phenylethane (6) in benzene at 105 °C, the results from which are quoted in Table 15.2. From the distribution of dimeric products formed from the coupling of the radical PhMeCH· within the solvent cage, and after diffusion from it, it is noted that (a) the yield of the *meso* dimer R—R′ is, for either reaction, equal to the total yield of the optically active dimers R—R and R′—R′, and (b) whereas random coupling after diffusion gives equal quantities of R—R and R′—R′, combination within the solvent cage leads to widely different quantities of these

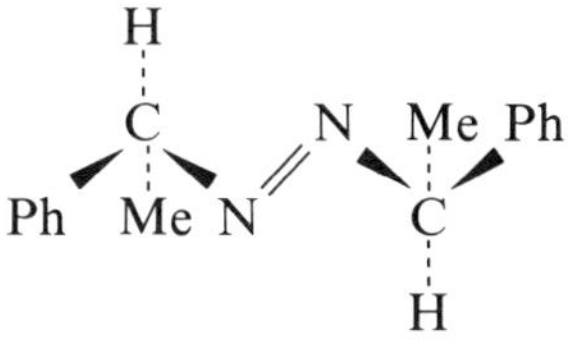

(6)

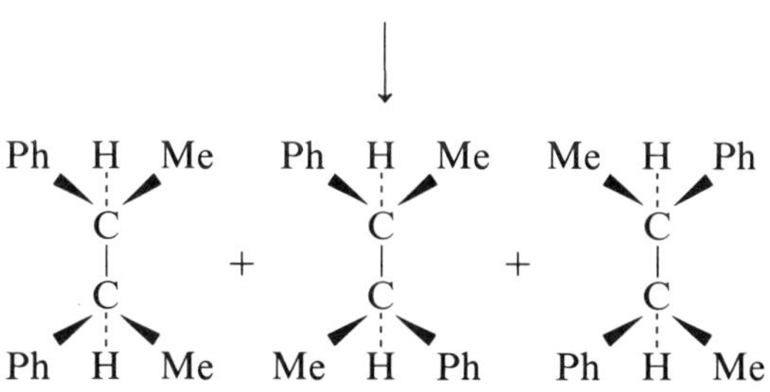

*meso*

(R—R′) (R—R) (R′—R′)

**Table 15.2** Distribution of Diastereoisomeric Products from Azo bis-α-phenylethane at 105 °C in Benzene

| Recombination Reaction | 2,3-Diphenylbutanes (% yield ± 1%) | | |
|---|---|---|---|
| | *Meso* | R—R | R′—R′ |
| Cage reaction | 48 | 21 | 31 |
| Reaction after diffusion | 50 | 25 | 25 |

isomers (21 % and 31 %). While the former observation is to be expected on statistical grounds, the latter suggests that considerable randomization of orientation of the radicals R· and R′·, relative to each other, does occur within the solvent cage prior to their combination.

## 15.5 Dimerization as a synthetic route

Kharasch[4] was the first to show the potential of this reaction as a synthetic method when, in 1942, he decomposed acetyl peroxide in acetic acid and isobutyric acid and obtained, respectively, succinic acid and tetramethylsuccinic acid in good yields. Aliphatic ketones furnished 1,4-diketones, as shown below for butanone:

$$\mathrm{MeCH_2COMe} \xrightarrow{\mathrm{Me\cdot}} \mathrm{Me\dot{C}HCOMe} \longrightarrow [\mathrm{MeCHCOMe}]_2$$

The dimethyl ether (7) of hexoestrol, a synthetic oestrogen, may similarly be synthesized, as follows:

$$\mathrm{MeO{-}C_6H_4{-}CH_2Et} \longrightarrow \mathrm{MeO{-}C_6H_4{-}\dot{C}HEt} \longrightarrow [\mathrm{MeO{-}C_6H_4{-}CHEt}]_2$$

(7)

Dimerization of radicals has also found application in the synthesis[9] of a series of potential oestrogens (8), which are difficult of access by other synthetic methods.

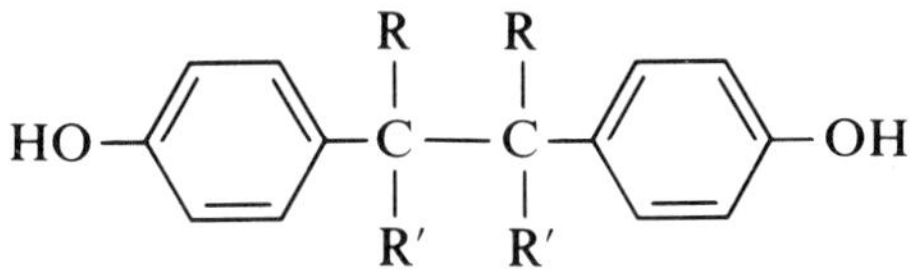

R = R′ = Me, Et, $Pr^n$
R = Et; R′ = Me or CN
R = $CO_2H$; R′ = H, Me, Et

(8)

Another interesting application is the preparation of the dimer (9) from tropolone, reported recently.[10]

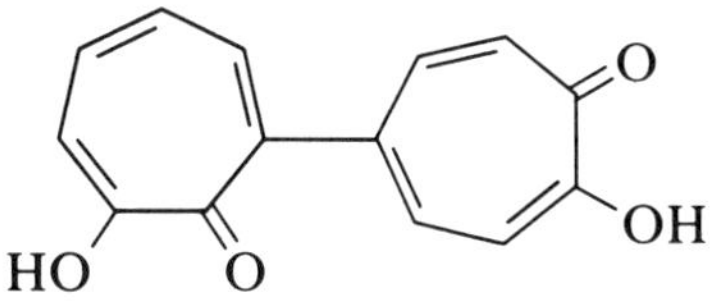

(9)

The coupling of phenoxy radicals, which are formed readily from the phenols by the action of oxidizing agents such as the ferricyanide ion, has great synthetic and biosynthetic significance.[11] Self-coupling produces dimers through formation of C—C, C—O, or O—O linkages, as illustrated below:

(a) Carbon-carbon Coupling:

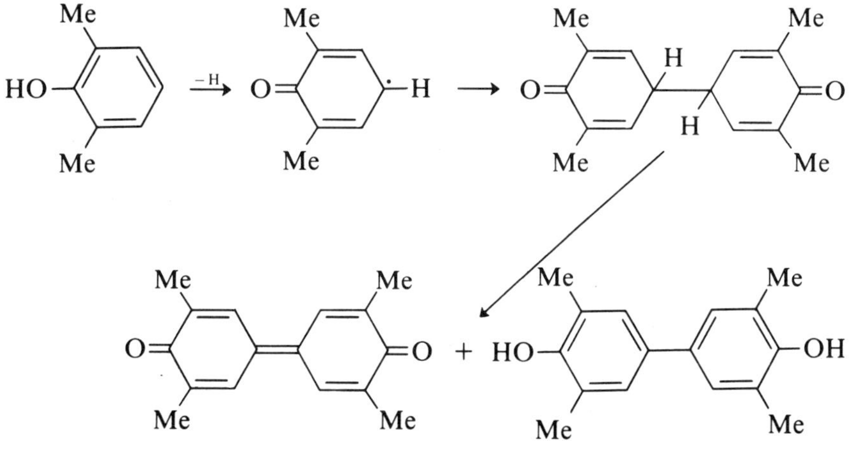

(b) Carbon-oxygen Coupling. The important antibiotic fungal metabolite, griseofulvin, was obtained in the racemic form by ferricyanide oxidation of the phenol (10) and subsequent reduction of the coupled product:

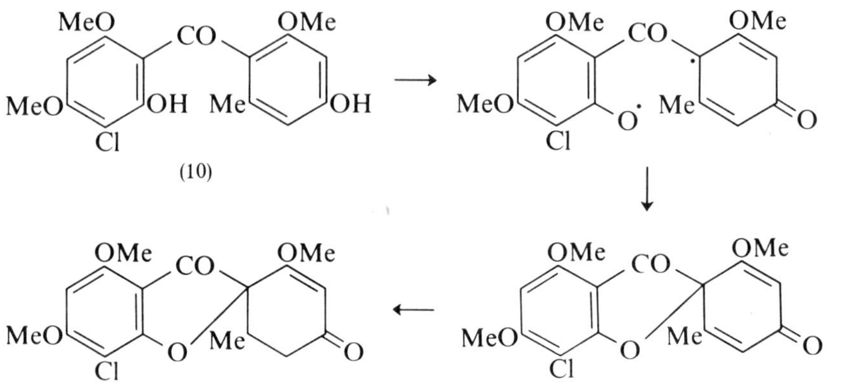

(c) Oxygen-oxygen Coupling. This is very rare as the O—O bond is generally much weaker than either the C—C or C—O bonds. With highly hindered phenoxy radicals, such as the 2,4,6-tri-*t*-butylphenoxy radical, coupling is sterically prevented and combination with small radicals such as oxygen may then occur:

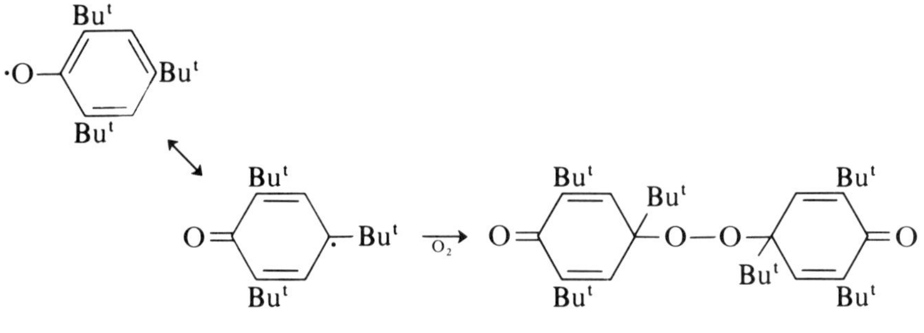

## References


1 P. D. Bartlett and J. M. McBride: Configuration, Conformation and Spin in Radical Pairs in *International Symposium on Free Radicals in Solution*, Butterworth, London, 1967, p. 89.

2 H. Lankamp, W. T. Nauta, and C. MacLean, *Tetrahedron Lett.*, 1968, 249; W. B. Smith, *J. chem. Educ.*, 1970, **47**, 535.

3 A. F. Trotman-Dickenson, *Free Radicals*, Methuen, London, 1959, p. 53; J. Grotewold and J. A. Kerr, *J. chem. Soc.*, 1963, 4337, 4342.

4 M. S. Kharasch and M. T. Gladstone, *J. Am. chem. Soc.*, 1943, **65**, 15.

5 M. S. Kharasch, H. C. McBay, and W. H. Urry, *J. org. Chem.*, 1945, **10**, 394, 401.

6 R. L. Huang and S. Singh, *J. chem. Soc.*, 1958, 891.

7 S. H. GOH, S. H. ONG, and I. SIEH, *Org. Magn. Resonance*, 1971, **3**, 713.
8 F. D. GREENE, M. A. BERWICK, and J. C. STOWELL, *J. Am. chem. Soc.*, 1970, **92**, 867.
9 R. L. HUANG and F. MORSINGH, *J. chem. Soc.*, 1953, 160; R. L. HUANG and K. T. LEE, *J. chem. Soc.*, 1954, 2570; 1955, 4229.
10 K. DOI and N. CHIBA, *Tetrahedron Lett.*, 1971, 2891.
11 C. J. M. STIRLING, *Radicals in Organic Chemistry*, Oldbourne, London, 1965, Chapter 10.

# 16
# Radical Rearrangements

Compared to the wealth of information available on carbonium ion rearrangements, the literature on free radical rearrangements is scanty. Despite the fact, however, that radicals show much less tendency to undergo rearrangement than the more electron-deficient species, many types of radical rearrangements are known.[1] Of these the 1,2-halogen shift is the most widespread and the 1,2-phenyl shift is also common; the neophyl radical $PhCMe_2CH_2\cdot$ has earlier been noted to rearrange to the more stable tertiary radical $Me_2\dot{C}CH_2Ph$.

## 16.1 Bridged radicals

A truly intramolecular rearrangement of groups, as depicted in (1), may represent either a transition state or a high energy intermediate of short life. Simple molecular orbital theory[2] predicts that one molecular orbital of low

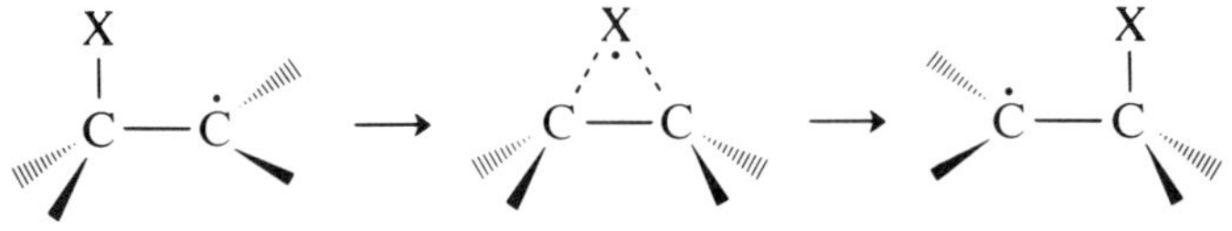

(1)

energy and two non-bonding orbitals of high energy are obtained from coupling of the *p* orbitals of two carbon atoms and one orbital of the migrating group. This is illustrated in Fig. 16.1. Thus a bridged carbonium ion will have its two electrons in the low energy level and consequently will be in an energetically favourable situation. In a bridged radical on the other hand the third electron must necessarily go into a high energy molecular orbital, with consequent destabilization. When it comes to a bridged carbanion, the two electrons in the high energy orbitals will give rise to a still more unfavourable energy situation. Furthermore, following the quantum mechanically derived Hunds rules a bridged carbanion would be in the triplet state. Thus it is understandable that carbonium ion rearrangements are widespread and indeed several bridged or 'non-classical' ions are known.

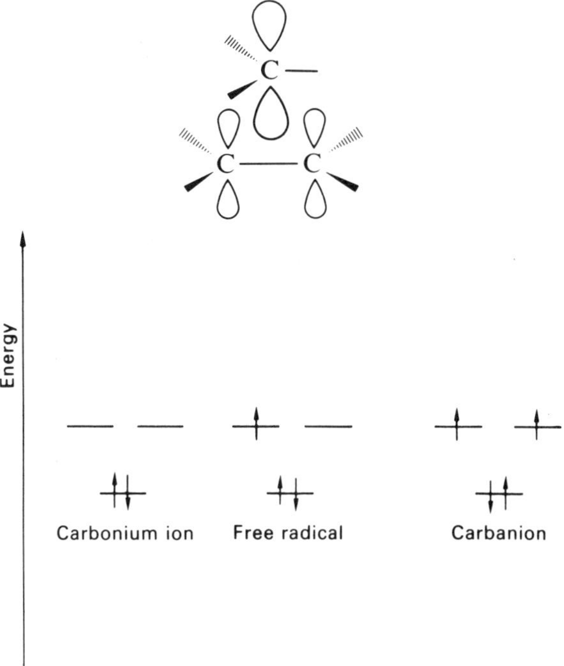


**Fig. 16.1** Energy levels of bridged species.

Radical rearrangements, however, would be expected to be limited in number since radical species of the same type would be somewhat destabilized. Finally, a bridged carbanion is an unlikely species and as a consequence truly 1,2-carbanion rearrangements have not been and perhaps never will be observed.

Although the occurrence of radical rearrangements involving halogen, aryl and sulphur groups are well established, much less is known as to whether the bridged structure (1) represents a transition state or an intermediate. Presently available evidence indicates that there are few such intermediates, bridged structures being usually high energy transition states in rearrangements. In the case of bridging bromine or iodine a symmetrical intermediate (2*a*) is likely, stabilization being possible here through delocalization of the unpaired electron into the *d* orbitals of the halogen. As to 1,2-alkyl or hydrogen shifts in radicals, these are unknown although aryl shifts are possible. The latter rearrangements do not proceed via bridged intermediates but delocalization and hence stabilization is possible in the transition state (2*b*). A symmetrical sulphur bridged radical was disproved by e.s.r. studies[3] but an unsymmetrical radical (2*c*) in which the sulphur atom interacts

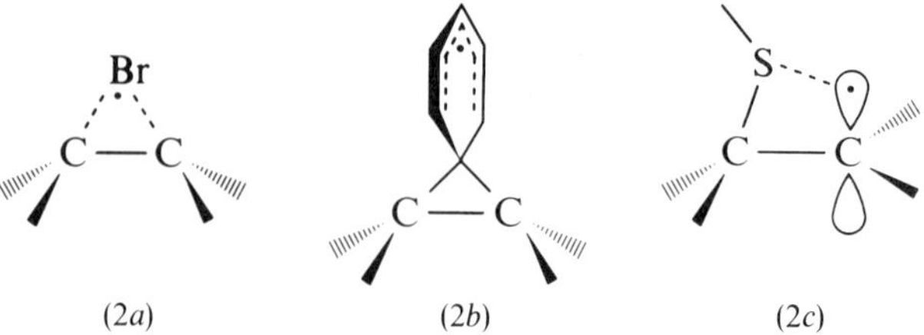


(2*a*) (2*b*) (2*c*)

sufficiently to inhibit bond rotation has been shown to be possible. Likewise the $\beta$-choroethyl radical prefers an unsymmetrical conformation.

## 16.2 1,2-Migrations of aryl groups

The rearrangement of the neophyl radical $PhCMe_2CH_2\cdot$ was first observed in 1944 by Kharasch and Urry[4] who prepared the radical by allowing 1-chloro-2-phenyl-2-methylpropane (neophyl chloride) to react with a Grignard reagent in the presence of cobaltous chloride. More illuminating information was later provided by Ruchardt[5] who generated the radical by the decarbonylation of $\beta$-phenylisovaleraldehyde (3) and by the decomposition of the perester (4). It was found that when decarbonylation was carried out in the absence of solvents the rearranged product, isobutylbenzene, was obtained

$$PhCMe_2CH_2CHO \ (3) \xrightarrow[-CO]{Bu^tO\cdot} \quad PhCMe_2CH_2CO_2OBu^t \ (4) \xrightarrow[-CO_2]{-Bu^tO\cdot}$$

$$PhCMe_2CH_2\cdot \rightarrow Me_2\dot{C}CH_2Ph$$

$$PhCMe_2CH_2\cdot \downarrow PhCMe_2CH_3 \qquad Me_2\dot{C}CH_2Ph \downarrow Me_2CHCH_2Ph$$

in a yield of 60%. This rose to 80% when chlorobenzene was used as solvent but in the presence of benzyl mercaptan, an effective hydrogen donor, dropped to less than 2%. These results can be explained as due to a competition between the rearrangement and the transfer reaction. In the presence of chlorobenzene as a diluent, the concentration of the transfer agent, i.e., the aldehyde, is reduced and hence rearrangement predominates. In the presence of an effective hydrogen donor practically all the radicals are scavenged before rearrangement can occur. These reactions are depicted below:

$$PhCMe_2CH_2\dot{C}O \rightarrow PhCMe_2CH_2\cdot \begin{cases} \xrightarrow{PhCMe_2CH_2CHO} PhCMe_3 + PhCMe_2CH_2\dot{C}O \\ \longrightarrow Me_2\dot{C}CH_2Ph \rightarrow Me_2CHCH_2Ph \\ \xrightarrow{PhCH_2SH} PhCMe_3 + PhCH_2S\cdot \end{cases}$$

Experiments designed to determine whether a bridged radical intermediate (2*b*), a species analogous to the phenonium ion in carbonium ion chemistry, is formed during phenyl migration revealed that such an intermediate is unlikely.[5] Thus the decarbonylation of optically active 3-methyl-3-phenylvaleraldehyde (5) gave only an inactive rearranged product. Further, the kinetics of decomposition of a variety of peresters (6) do not show enhanced rates required of aryl participation. Thus it is quite clear the aryl bridged radicals are not intermediates in these reactions. The driving force for aryl migration appears to be derived from the formation of a more stable tertiary radical.

$$\mathrm{Ph{-}C(Me)(Et){-}CH_2\dot{C}O} \;(5) \nearrow \mathrm{Ph{-}C(Me)(Et){-}CH_2\cdot} \rightarrow \mathrm{Me(Et)\dot{C}{-}CH_2Ph} \rightarrow \mathrm{Me(Et)HC{-}CH_2Ph}$$

(5) $\not\searrow$ bridged radical (C—C)

$$\mathrm{Ar{-}C(R_1)(R_2){-}CH_2CO_2OBu^t} \;(6) \searrow \mathrm{Ar{-}C(R_1)(R_2){-}CH_2\cdot + CO_2 + Bu^tO\cdot}$$

(6) $\not\nearrow$ bridged radical

$R_1$ = H, Me, Ph
$R_2$ = H, Me, Ph
$R_3$ = Ph, *p*-$MeOC_6H_4$, *p*-$ClC_6H_4$, 9-anthranyl

Available evidence also argues against alkyl or hydrogen bridged radicals. In fact, an authentic instance of 1,2-alkyl (or hydrogen) migration in a hydrocarbon monoradical has yet to be demonstrated. The rate of decomposition of the *exo* isomer of the perester (7) is only three times faster than that of the *endo* isomer. Thus the 2-norbornyl radical is not 'non-classical' and likewise the existence of other carbon bridged non-classical radicals structurally analogous to the better known non-classical cations have yet to be unambiguously demonstrated. For instance e.s.r. studies show that the 7-norbornyl radical (8) is classical rather than non-classical, the radical centre being pyramidal with the 7-hydrogen bent away from the double bond.[6]

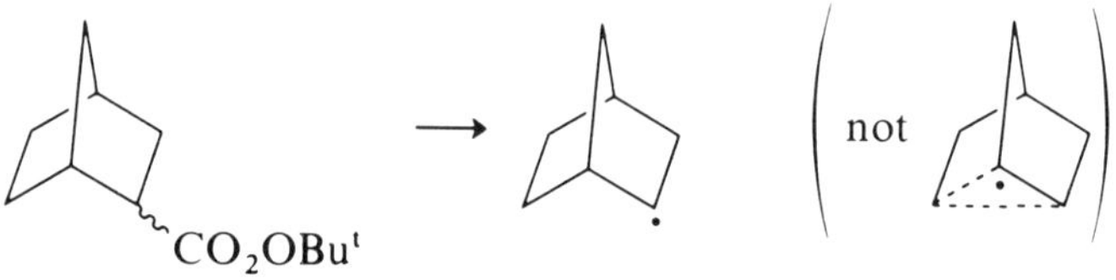

$CO_2OBu^t$

(7)

not

H

not

H

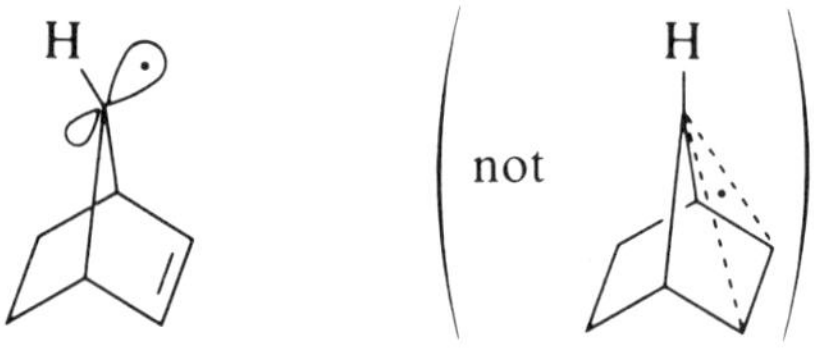

(8)

### 16.3 Halogen migrations

By far the best documented of radical rearrangements is the halogen shift in polyhalogenalkyl radicals.[1a] 1,2-Chlorine shifts appear to be particularly facile. In the simplest instance, the homolytic addition of hydrogen bromide to 3,3,3-trichloropropene gives 1,1,2-trichloro-3-bromopropane in high yield. The reaction is considered to involve an intramolecular chlorine shift rather than an addition–elimination–readdition sequence. This is supported by the fact that in most cases of chlorine rearrangements no side product

$$CCl_3CH{=}CH_2 \xrightarrow{Br\cdot} CCl_3\dot{C}HCH_2Br \longrightarrow \dot{C}Cl_2CHClCH_2Br$$

$$CCl_2{=}CHCH_2Br + Cl\cdot$$

$$\dot{C}Cl_2CHCH_2Br \xrightarrow{HBr} HCCl_2CHClCH_2Br + Br\cdot$$

arising from the possible intermolecular capture of any eliminated chlorine atom could be found. Addition–elimination however, has been known to occur in rare instances, e.g., the homolytic addition of thiophenol to 3,3,3-trichloropropylene gives a small yield of hydrogen chloride and an olefin. 1,2-Bromine shifts are less common, as the bromine atom $\alpha$ to the radical

$$CCl_3CH{=}CH_2 + PhS\cdot \rightarrow CCl_3\dot{C}HCH_2SPh$$

$$CCl_3\dot{C}HCH_2SPh \begin{cases} \nearrow \dot{C}Cl_2CHClCH_2SPh \rightarrow HCCl_2CHClCH_2SPh \\ \searrow CCl_2{=}CHCH_2SPh + Cl\cdot \rightarrow HCl \end{cases}$$

centre tends to be eliminated giving rise to an olefinic product. An $\alpha$ bromine appears to have a stabilizing effect on the radical and indeed Skell[7] has proposed that bridged radicals are actually formed. For instance it has been found that homolytic bromination (by molecular bromine, *N*-bromosuccinimide or *t*-butoxy bromide) of optically active 1-bromo-2-methylbutane gives optically active 1,2-dibromo-2-methylbutane, a result explained as due to the intermediacy of a bridged radical.[9] This intermediate is probably not particularly stable since at concentrations of less than 0.05*M* bromine, ring opening which gives a racemic product begins to compete with the transfer reaction. Bromination of the analogous optically active 1-chloro-2-methylbutane also leads to an optically active product. However, in this

$$(Et)(Me)(H)C{-}CH_2Br \xrightarrow{Br\cdot} (Et)(Me)C\overset{\dot{Br}}{\frown}CH_2\ (9) \xrightarrow{Br_2} (Et)(Me)(Br)C{-}CH_2Br + Br\cdot$$

$$(9) \rightleftharpoons (Et)(Me)\dot{C}{-}CH_2Br \rightarrow dl\ \text{product}$$

case large amounts of racemization occurs even at high concentrations of bromine. The greater ability of bromine to form bridged species is also demonstrated in the photobromination of straight chain aliphatic bromides (but not the chlorides) which show a great tendency to give selectively high yields of *vicinal* dibromides. Thus an 85% yield of 1,2-dibromobutane is obtained by the photobromination of *n*-butyl bromide in contrast to a mere 25% of 2-bromo-1-chlorobutane from *n*-butyl chloride.

The ability of bromine to form radical intermediates is further demonstrated in the photobromination of substituted cyclohexanes.[8] For example, significant differences have been observed in the reaction pattern of the photobromination of the isomers of 4-*t*-butylcyclohexyl bromides. The *cis* isomer (10*a*) gives a 90% yield of the *trans* dibromide stereospecifically with no product arising from the tertiary carbon–hydrogen bonds, while the *trans* isomer (10*b*), in contrast, yields a mixture of products arising from substitution at various positions in the ring. Furthermore a competitive bromination reaction of the two isomers showed that the *cis* isomer is at least fifteen times more reactive than the *trans* compound. Since the rate of photobromination of the *trans* isomer appears normal the enhanced rate of the *cis* compound must be due to neighbouring group participation (anchimeric assistance) in the transition state (11). It may be pointed out that the substantial assistance observed here may partly be due to the polarized nature of the transition state (12), the developing cationic centre of which can be stabilized by the

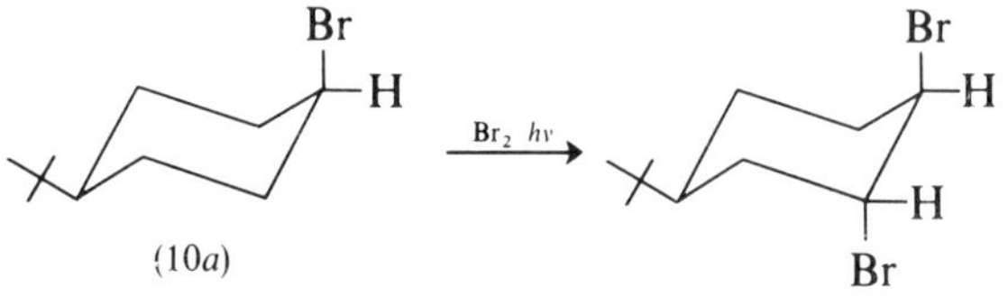

Br $\xrightarrow{Br_2\ h\nu}$ 90%

(10*a*)

→ mixture of products

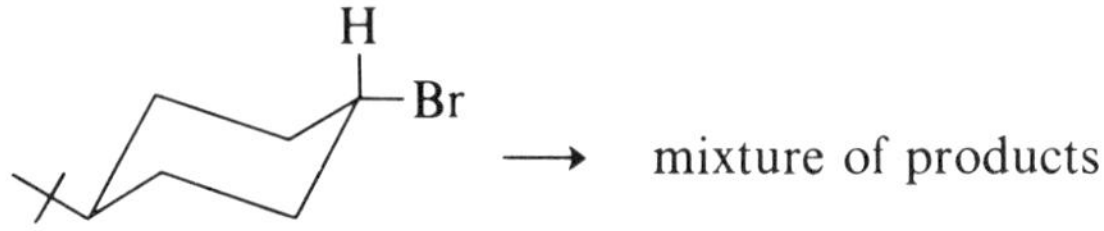

(10*b*)

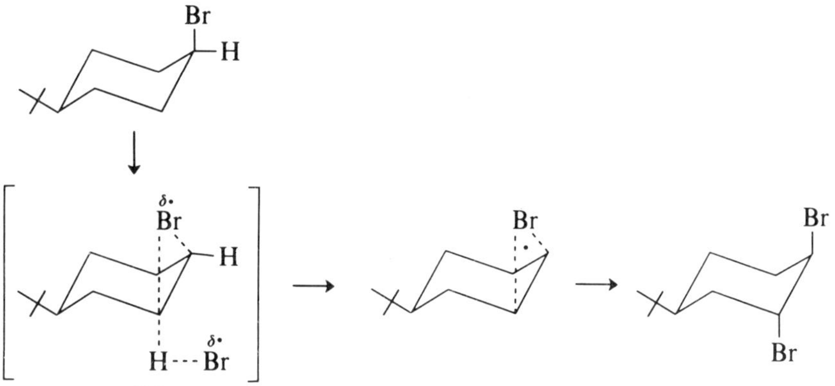

(11)

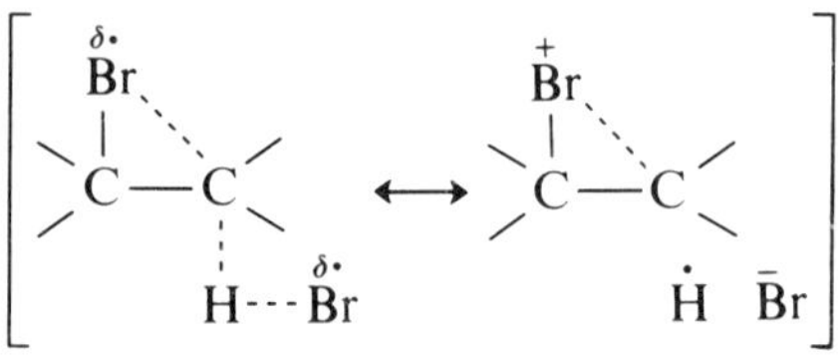

(12)

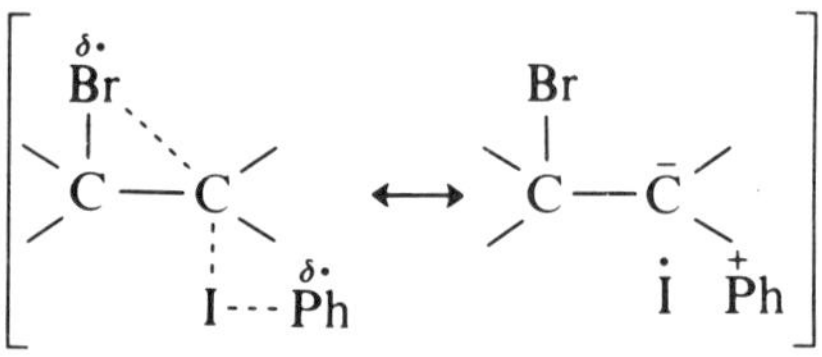

(13)

neighbouring bromine group. In the iodine abstraction from 1-bromo-2-iodoethane by phenyl radicals, a case in which a differently polarized transition state (13) may be expected, a small rate enhancement has also been observed.[9] This is again indication that participation by neighbouring bromine is possible.

Electron spin resonance data on radicals formed by the addition of atomic bromine to olefins in a matrix are in agreement with the existence of bridged bromine radicals.[10] It is quite evident that bromine bridging confers a certain amount of stabilization to the odd electron. If this is due to the availability of the low lying *d* orbitals, then it may be expected that iodine bridging should also be observed. Few experiments on this have been reported.[7] The photo-initiated additions of iodine to *cis* and *trans* butenes are *trans* stereospecific. These reactions remain stereospecific at very low concentrations ($10^{-5}M$) of iodine, indicating the remarkable stability of the bridged radical. Ring opening of the bridged radical would have led to non-stereospecific addition as has sometimes been observed in bromination as a result of the relatively less stable bromine bridged radical (see Section 12.4).

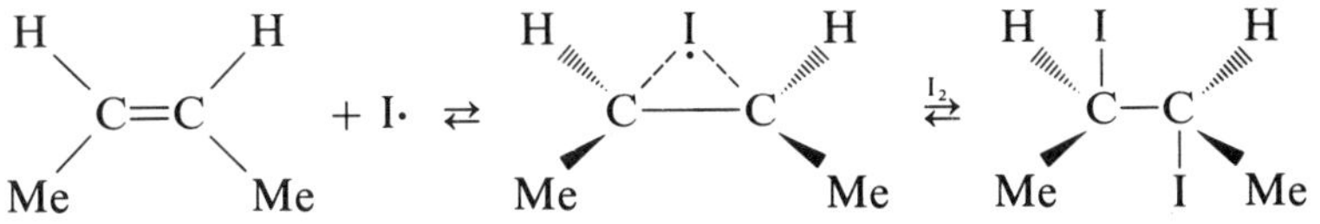

H(Me)C=C(H)Me + I· ⇄ [bridged iodine radical: H, Me-C—C-H, Me with I·] $\overset{I_2}{\rightleftarrows}$ H(I)(Me)C—C(H)(I)Me

*dl*-2,3-diiodobutane

## 16.4 Homoallylic radical rearrangements

While alkyl group migrations rarely, if ever, take place in free radicals, the 1,2-shifts of vinyl groups do occur. This rearrangement is structurally analogous to the rearrangement of the neophyl radical discussed previously.

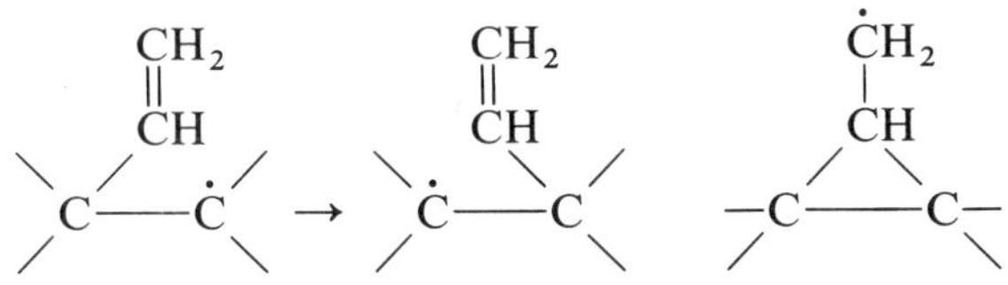

$CH_2{=}CH{-}C{-}\dot{C} \rightarrow \dot{C}{-}C{-}CH{=}CH_2$ ; cyclopropylcarbinyl radical $\dot{C}H_2{-}CH<(C{-}C)$

(14)

Montgomery and co-workers[11] have demonstrated that the reaction takes place via the intermediacy of a cyclopropylcarbinyl radical (14). The *t*-butyl peroxide initiated decarbonylation of 3-methyl-methyl-4-pentenal has been found to give 3-methyl-1-butene and 1-pentene, the yield of the latter increasing with decreasing concentration of the aldehyde. Traces of 1,2-dimethyl cyclopropanes can also be detected by gas chromatography but generally such cyclized products are not obtained from this rearrangement

$$CHO{-}CH_2{-}CH(CH_3){-}CH{=}CH_2 \xrightarrow{Bu^tO\cdot} \dot{C}H_2{-}CH(CH_3){-}CH{=}CH_2 \rightleftarrows \overset{\cdot}{C}H_2{-}\underset{\text{cyclopropane: }CH_2{-}CH{-}CH(CH_3)}{} \rightleftarrows CH_2{=}CH{-}CH_2{-}\dot{C}H{-}CH_3$$

$$\dot{C}H_2{-}CH(CH_3){-}CH{=}CH_2 \rightarrow (CH_3)_2CH{-}CH{=}CH_2 \qquad CH_2{=}CH{-}CH_2{-}\dot{C}H{-}CH_3 \rightarrow CH_3CH_2CH_2CH{=}CH_2$$

except when the cyclopropylcarbinyl radical is sufficiently stabilized as in the 1,1-diphenylcyclopropylcarbinyl radical. It may be noted that the ring opening of the cyclopropylcarbinyl radical has been found to be a very rapid process, comparable to molecular diffusion.

$$CHO{-}CH_2{-}CH_2{-}CH{=}CHD \;(\text{D and H } cis) \rightarrow H_2\dot{C}{-}CH_2{-}CH{=}CHD \;(cis) \rightleftarrows H_2C{-}CH_2{-}CH\text{(ring)}{-}\dot{C}HD \;(15) \rightleftarrows H_2C{=}CH{-}CH_2\cdot \text{ with } CHD \;(trans)$$

$$H_2\dot{C}{-}CH_2{-}CH{=}CHD \rightarrow CH_3{-}CH_2{-}CH{=}CHD \;(cis) \qquad \rightarrow CH_3{-}CH_2{-}CH{=}CHD \;(trans)$$

*cis* (15) *trans*

Further evidence[11] for the formation of a cyclopropylcarbinyl radical intermediate comes from a study of the decarbonylation of *cis* 4-pentenal-5-$d_1$ which gives *cis* 1-butene-1-$d_1$ and *trans* 1-butene-$d_1$. At low concentrations of the aldehyde (1*M*) approximately equal amounts of *cis* and *trans* olefins result but decarbonylation of the neat aldehyde (6.4*M*) leads to a ratio of *cis* to *trans* olefin of 1.42. This dependence on concentration is found presumably because the trapping of unrearranged radical occurs at a rate comparable to that of rearrangement. These results show that the cyclopropylcarbinyl radical (15) is an intermediate and not merely a transition state, since a loss of geometrical identity means that bond rotation must have taken place as shown (transition state lifetimes are generally considered short relative to internal rotational lifetimes).

## 16.5 1,2-Rearrangements to carbanionic centres

Meisenheimer, Stevens, and Wittig rearrangements,[12] exemplified by the equations below, all involve the net 1,2-migration of alkyl groups to carbanionic centres. These rearrangements are intramolecular and an optically active migrating group may partially lose its configuration on migration, as in Wittig and Meisenheimer rearrangements, or may completely retain its configuration, as frequently observed in the Stevens rearrangement. This

$$\mathrm{PhCOCH_2\overset{+}{S}(CH_2Ph)MeBr^-} \xrightarrow{\text{base}} \mathrm{PhCO\overset{-}{C}H\overset{+}{S}(CH_2Ph)Me} \longrightarrow \mathrm{PhCOCH(CH_2Ph)SMe} \quad \text{(Stevens)}$$

$$\mathrm{PhCOCH_2\overset{+}{N}(CH_2Ph)Me_2} \xrightarrow{\text{base}} \mathrm{PhCO\overset{-}{C}H\overset{+}{N}(CH_2Ph)Me_2} \longrightarrow \mathrm{PhCOCH(CH_2Ph)NMe_2} \quad \text{(Stevens)}$$

$$\mathrm{PhCH_2OMe} \xrightarrow{\text{base}} \mathrm{Ph\overset{-}{C}HOMe} \longrightarrow \mathrm{PhCH(Me)O^-} \quad \text{(Wittig)}$$

$$\mathrm{Ph{-}N^+(Me)(CH_2Ph){-}O^-} \longrightarrow \mathrm{PhN(Me){-}OCH_2Ph} \quad \text{(Meisenheimer)}$$

and other evidence points to the fact that the rearrangement proceeds by cleavage to a short-lived radical pair which recombines to give the product. A concerted mechanism with retention of configuration of the migrating group is unlikely while a carbanion bridged species is objectionable on theoretical grounds.

$$PhCO\overset{-}{C}H\overset{+}{N}(CH_2Ph)Me_2 \rightarrow \left[PhCO\overset{-}{C}H\overset{+}{\dot{N}}Me_2 \leftrightarrow PhCO\dot{C}HNMe_2 \quad \cdot CH_2Ph\right] \rightarrow PhCOCH(CH_2Ph)NMe_2$$

$$PhCO\overset{-}{C}H\overset{+}{S}(CH_2Ph)Me \rightarrow \left[PhCO\overset{-}{C}H\overset{+}{\dot{S}}Me \leftrightarrow PhCO\dot{C}HSMe \quad \cdot CH_2Ph\right] \rightarrow PhCOCH(CH_2Ph)SMe$$

$$Ph{-}\overset{Me}{\underset{CH_2Ph}{N^+}}{-}O^- \rightarrow \left[Ph{-}\overset{Me}{\dot{N}^+}{-}O^- \leftrightarrow Ph{-}\overset{Me}{N}{-}O\cdot \quad \cdot CH_2Ph\right] \rightarrow Ph{-}\overset{Me}{N}{-}OCH_2Ph$$

$$Ph\overset{-}{C}HOR \rightarrow \left[Ph\overset{-}{C}HO\cdot \leftrightarrow Ph\dot{C}HO^- \quad R\cdot\right] \rightarrow Ph\underset{R}{C}HO^-$$

### 16.6 Radical chain 1,3-migrations

The large group of rearrangements of enol ethers and other structurally related compounds may be represented by the following equations.[13]

$$Ph{-}\overset{OMe}{C}{=}CH_2 \rightarrow Ph\overset{O}{\overset{\|}{C}}CH_2CH_3$$

$$Me{-}\overset{OTs}{C}{=}CH_2 \rightarrow Me{-}\overset{O}{\overset{\|}{C}}{-}CH_2Ts, \qquad Ts = p\text{-}MeC_6H_4SO_2$$

$$H{-}\overset{OMe}{C}{=}N{-}Ph \rightarrow H{-}\overset{O}{\overset{\|}{C}}{-}N(Me)(Ph)$$

Their radical chain nature is supported by a number of observations: (i) the reactions are catalysed by peroxides and related initiators and slowed down by inhibitors, (ii) optically active migrating groups undergo complete racemization, and (iii) in the presence of compounds containing easily abstractable hydrogen atoms, the migrating alkyl group forms an alkane while other radical fragments are incorporated into the reaction product. A mechanism consistent with these observations is the following:

$$\mathrm{Ph{-}C(OR){=}CH_2 \rightarrow Ph{-}C(O\cdot){=}CH_2 + R\cdot}$$

$$\mathrm{R\cdot + Ph{-}C(OR){=}CH_2 \rightarrow Ph{-}\dot{C}(OR){-}CH_2R \rightarrow Ph{-}C({=}O){-}CH_2R + R\cdot}$$

## 16.7 Miscellaneous rearrangements

Under this heading only a few selected cases will be discussed. As rearrangements are not as important or widespread in radical chemistry as they are in carbonium ion chemistry, the examples given would be more appropriately regarded as exceptions rather than the rule in radical reactions. Intramolecular hydrogen abstractions, sometimes treated as 1,5-hydrogen shifts, have been mentioned earlier and will not be discussed here, nor will various fragmentation processes (Chapter 14) which occur accompanied by structural changes and isomerization.

The first example of an 1,2-acyloxy migration[14] was found in the decarbonylation of 3-acyloxy-3-methylbutanal, to give *t*-butyl acetate and isobutyl acetate. The 1,2-shift probably occurs via the transition state (16)

$$\mathrm{Me{-}C(Me)(OCOMe){-}CH_2CHO \xrightarrow{(PhCOO)_2} Me{-}C(Me)(OCOMe){-}CH_2\cdot \rightarrow Me{-}C(Me)(O{\cdots}\dot{C}(Me){\cdots}O){-}CH_2\ (16)}$$

$$\mathrm{Me{-}C(Me)(OCOMe){-}CH_2\cdot \rightarrow Me{-}C(Me)(Me){-}OCOMe}$$

$$\mathrm{(16) \rightarrow Me{-}\dot{C}(Me){-}CH_2OCOMe \rightarrow Me_2CHCH_2OCOMe}$$

but more information will be needed before this can be settled. The driving force behind the reaction is presumably derived from the transformation of a primary radical to a more stable tertiary radical.

An example of a 1,5-phenyl migration is provided by the reaction of phenylbutyldimethylsilane with *t*-butyl peroxide.[15] The rearranged product was obtained in 31% yield together with 69% of a cyclization product. The ring size of the 6-membered *spiro* intermediate (17) appears to be of special importance in this rearrangement since no rearrangement was observed for other homologous radicals of the same series.

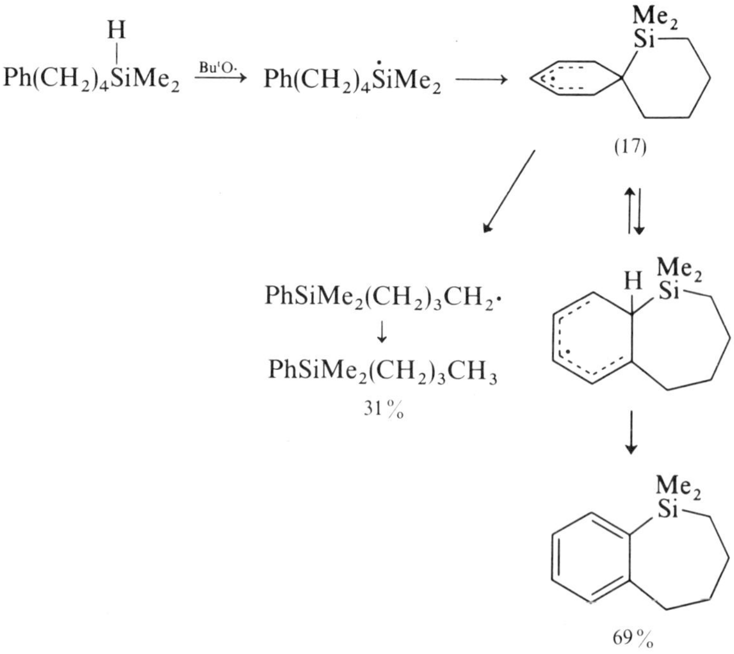

$$Ph(CH_2)_4SiHMe_2 \xrightarrow{Bu^tO\cdot} Ph(CH_2)_4\dot{Si}Me_2 \longrightarrow (17)$$

(17)

$$PhSiMe_2(CH_2)_3CH_2\cdot \downarrow PhSiMe_2(CH_2)_3CH_3 \quad 31\%$$

69%

1,4-Phenyl migrations of similar mechanism have also been observed in the decarbonylation of 5-methyl-5-phenylhexanal and in the electrolysis of $\beta,\beta,\beta$-triphenylpropionic acid.[1b]

## References


1a R. KH. FREIDLINA, *Adv. Free-Radical Chem.*, 1965, **1**, Chapter 6, G. H. WILLIAMS, Ed., Logos, London.

b C. WALLING, *Molecular Rearrangements*, P. DE MAYO, Ed., Interscience, New York, 1963, Part 1, Chapter 7.

c L. KAPLAN, *Bridged Free Radicals*, Marcel Dekker, New York, 1972.

2 A. STREITWIESER, *Molecular Orbital Theory for Organic Chemists*, Wiley, New York, 1961, p. 380.

3 P. J. KRUSIC and J. K. KOCHI, *J. Am. chem. Soc.*, 1971, **93**, 846.
4 M. S. KHARASCH and W. H. URRY, *J. Am. chem. Soc.*, 1944, **66**, 1438.
5*a* C. RUCHARDT and R. HECHT, *Chem. Ber.*, 1965, **98**, 2460, 2471.
*b* C. RUCHARDT and H. TRAUTWEIN, *Chem. Ber.*, 1965, **98**, 2478.
6 P. BAKUZIS, J. K. KOCHI, and P. J. KRUSIC, *J. Am. chem. Soc.*, 1970, **92**, 1434.
7 P. S. SKELL, *Organic Reaction Mechanisms*, Special Publication, No. 19, The Chemical Society, London, 1965, p. 131.
8 P. S. SKELL and P. D. READIO, *J. Am. chem. Soc.*, 1964, **86**, 3334.
9 W. C. DANEN and R. L. WINTER, *J. Am. chem. Soc.*, 1971, **93**, 716.
10 P. I. ABELL and L. H. PIETTE, *J. Am. chem. Soc.*, 1962, **84**, 916.
11 L. K. MONTGOMERY and J. W. MATT, *J. Am. chem. Soc.*, 1967, **89**, 934, 3050, 6556.
12*a* U. SCHOLLKOPF, *Angew. Chem., Int. Ed.*, 1970, **9**, 763.
*b* R. A. W. JOHNSTONE, *Mechanisms of Molecular Migrations*, B. S. THYAGARAJAN, Ed., Wiley, 1969, Vol. 2, p. 249.
*c* A. R. LEPLEY, P. M. COOK, and G. F. WILLARD, *J. Am. chem. Soc.*, 1970, **92**, 1101.
13 P. S. LANDIS, *Mechanisms of Molecular Migrations*, B. S. THYAGARAJAN, Ed., Wiley, 1969, Vol. 2, p. 43.
14 D. D. TANNER and F. C. P. LAW, *J. Am. chem. Soc.*, 1969, **91**, 7535.
15 H. SAKURAI and A. HOSOMI, *J. Am. chem. Soc.*, 1970, **92**, 7507.

# 17

# Carbenes

## 17.1 Structure of carbenes

Since serious study of carbene chemistry began two decades ago, development of this field has been exceptionally rapid, owing partly to the synthetic utility of carbenes in the preparation of cyclopropanes and no less to the growing interest in the mechanism of carbene reactions.[1] Carbenes, or methylenes, may be defined as divalent carbon intermediates having two covalent bonds and two non-bonding electrons. Common examples of carbene intermediates are methylene ($CH_2$:), chlorocarbene (ClCH:), phenylcarbene (Ph—CH:), diphenylcarbene ($Ph_2C$:), and ethoxycarbonylcarbene ($EtO_2C$—CH:). With two non-bonding electrons the carbene can be in the singlet state (anti-parallel spins) or the triplet state (parallel spins). Triplet carbenes have sometimes been referred to as diradicals although the location of two unpaired spins on the same carbon may give rise to unique chemical characteristics.

The structure of the simplest carbene, methylene, has been the object of considerable research. Theoretical calculations with various degrees of sophistication have appeared. Singlet methylene may be pictured as having $sp^2$ hybridization and a vacant $p$ orbital (1). The presence of a vacant $p$ orbital shows that singlet carbenes are electron-deficient species similar to carbonium ions, but on the other hand the presence of a non-bonding pair of electrons may give it the characteristics of carbanions. Triplet methylene may be described as having $sp$ hybridization, the two electrons with parallel spins occupying the two $p$ orbitals (2). The electronic spectra of the lowest singlet and triplet states of methylene were finally recorded by Herzberg[2] in 1959 after a 17 year search. The singlet state has a bent structure ($H\hat{C}H$ 103°) and is of higher energy than the triplet state which is approximately linear. Triplet carbenes need not be linear and this is shown in electron spin resonance studies[3] on triplet diphenylcarbene (3) and triplet fluorenylidene (4). Both carbenes have triplet ground states and are stable for several hours at 77 K when generated in a solvent glass. E.s.r. spectra of the two carbenes show that one electron is delocalized in the aromatic rings while the other

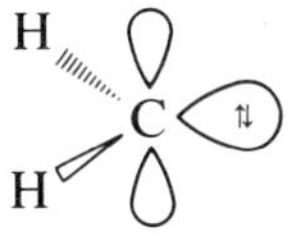

(1)

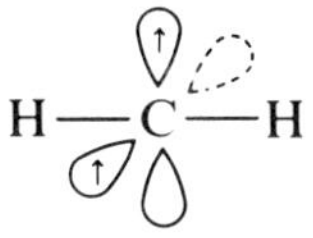

(2)

(3)

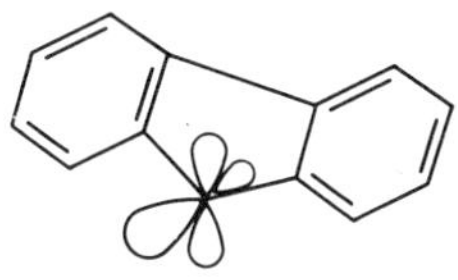

(4)

electron is localized on the carbene carbon. This would make the two triplet carbenes somewhat similar to the corresponding radicals.

Although all the carbenes described above have triplet ground states there are also carbenes with a singlet ground state, e.g., dihalocarbenes.[4] But, there being no unpaired electrons singlet carbenes are not accessible through e.s.r. studies.

## 17.2 Generation and reactions of carbenes

Carbenes can usually be generated by the photolysis or thermolysis of a variety of substances, especially diazo compounds, as indicated below:

$$CH_2{=}\overset{+}{N}{=}\overset{+}{N} \xrightarrow{h\nu \text{ or } \Delta H} CH_2 + N_2$$

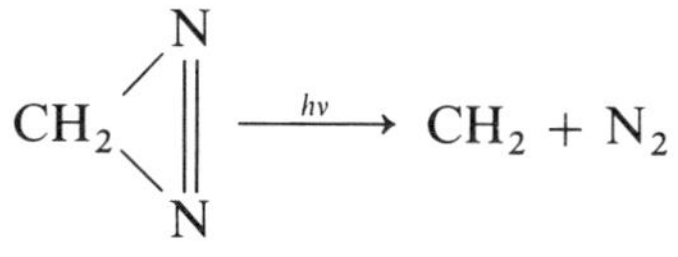

$$CH_2\langle\begin{smallmatrix}N\\ \|\\ N\end{smallmatrix} \xrightarrow{h\nu} CH_2 + N_2$$

$$CH_2{=}C{=}O \xrightarrow{h\nu} CH_2 + CO$$

The species obtained from these processes may be referred to as free carbenes since they are not associated with, or complexed to, other ions or molecules. Singlet carbenes are formed in most cases as a consequence of spin conservation unless the photolysis is carried out in the presence of a triplet sensitizer or unless a rapid intersystem crossing can take place. Thus triplet methylene can be produced by the photolysis of ketene sensitized by mercury atoms. The photolysis of diphenyldiazomethane and 9-diazofluorene give free carbenes in both the singlet and triplet states.

Two major types of reactions have been observed for carbenes namely, insertion into a $\sigma$-bond, and addition to an unsaturated bond. In the absence of substrates with which the carbenes generated can react readily, dimerization takes place. These reactions are shown below.

$$\rangle C{-}H + CH_2 \rightarrow \rangle C{-}CH_2{-}H \quad \text{(insertion)}$$

$$\rangle C{=}C\langle + CH_2 \rightarrow \rangle C \overset{}{\underset{CH_2}{\frown}} C\langle \quad \text{(addition)}$$

$$2CH_2 \rightarrow CH_2{=}CH_2 \quad \text{(dimerization)}$$

There is a large number of compounds and intermediates which may be considered as carbene fragments complexed or associated with metal ions with various degrees of covalent bonding. Among these are relatively stable α-halo-organomercurials, moderately stable α-halo-organozinc compounds and unstable α-halo-organolithiums. Because of the close resemblance of the reactivity of these species to that expected of free carbenes they have been appropriately referred to as carbenoids (Closs and Moss).[5] The simplest formulation of a carbenoid species may be written as (5), in which M is a cationic group, usually a metal, and X is an electronegative element or group.

$$RR'C(M)(X)$$

(5)

Carbenoid species are of considerable importance in the synthesis of cyclopropanes and highly strained small ring systems. α-Elimination of hydrogen halides by a strong base (Hine)[6] provides a convenient method for the

$$CHBr_3 \xrightarrow{t\text{-BuO}^-K^+} \bar{C}Br_3\,K^+ \left( \rightarrow Br_2C: + KBr \right) \longrightarrow$$

$$\xrightarrow[70\%]{\text{cyclohexa-1,3-diene}} \text{7,7-dibromobicyclo[4.1.0]hept-2-ene}$$

generation of carbenoid species. The use of alkyllithiums as base can bring about the α-elimination of halogen. Intramolecular carbene insertions or additions have been used to synthesize novel small ring compounds, two instances of which are shown:

Br Br — MeLi → Br Li → ( → :) → ... + ...

$CHN_2$ — $\Delta H$ → CH: → ...

A large number of catalysts are known which can bring about the decomposition of diazo compounds to give carbenoid intermediates. These catalysts include copper metal, copper salts and complexes, and several metal halides. The catalytic decomposition of diazo compounds by zinc halides leads to the formation of organozinc intermediates and these in the presence of olefins give good yields of cyclopropanes. In such additions to double bonds the zinc carbenoids display remarkable stereoselectivity. For instance, the thermodynamically less stable *syn* cyclopropane isomer is formed as a major product in the catalytic decomposition of aryldiazomethanes in the presence of olefins. The reaction of methylene iodide with a zinc/copper couple yields iodomethylzinc iodide or Simmons-Smith reagent,[7] which is widely used in the synthesis of cyclopropanes. The same intermediate is formed from the reaction of diazomethane with zinc iodide.

$$Ar{-}CHN_2 + ZnX_2 \rightarrow ArCHXZnX \longrightarrow$$ Ar + Ar

*syn* (major product) *anti*

$$CH_2I_2 + Zn/Cu \searrow$$
$$ICH_2ZnI \xrightarrow[60\%]{} $$
$$CH_2N_2 + ZnI_2 \nearrow$$

Dihalocarbenes have been generated from the thermolysis of α-halo-organomercurials.[8] This method has the advantage of the carbene being generated under relatively neutral conditions and thus allows base-sensitive olefins to be cyclopropanated. Poorly nucleophilic olefins can also react.

Tetrachloroethylene, for example, reacts with phenyl(trichloromethyl)-mercury to give a 74% yield of hexachlorocyclopropane.

Cyclopropene synthesis may be achieved via carbene intermediates. The synthesis of 1,3,3-trimethylcyclopropene is outlined below.

$$(CH_3)_2C{=}C(CH_3)CH{=}NNHSO_2Ar \xrightarrow{NaOMe} (CH_3)_2C{=}C(CH_3)CHN_2 \rightarrow (CH_3)_2C{=}C(CH_3)CH\!: \rightarrow \text{1,3,3-trimethylcyclopropene} \; (72\%)$$

## 17.3 Structure and reactivity of carbenes

The photolysis of diazomethane in the liquid phase produces free singlet methylene. This, because of the high frequency of collision with the solvent, usually reacts with it rather than decaying to the triplet state. In the gas phase, however, the singlet species can decay to the lower energy triplet state.

Free singlet methylene is a very reactive intermediate and inserts into C—H bonds indiscriminately. It is partly for this reason that methylene has become known as one of the most 'indiscriminate' reagents in organic chemistry (Doering *et al.*)[9] The liquid phase photolysis of diazomethane in 2-methylbutane leads to almost random, statistical insertion by the carbene, the attacking entity hardly discriminating among tertiary, secondary, and primary C—H bonds. A comparison of the reactivity towards these three

$$CH_3{-}CH(CH_3){-}CH_2CH_3 + CH_2 \rightarrow \underset{1.0}{H{-}CH_2{-}CH_2{-}CH(CH_3){-}CH_2CH_3} + \underset{1.05}{CH_3{-}CH(CH_3){-}CH_2CH_2{-}CH_2{-}H}$$

$$+ \underset{1.22}{CH_3{-}CH(CH_3){-}CH(CH_2{-}H)CH_3} + \underset{1.51}{CH_3{-}C(CH_3)(CH_2{-}H){-}CH_2CH_3}$$

**Table 17.1** Comparison of the Reactivities of Some Intermediates

| Reaction | Selectivity 1° | 2° | 3° | Stereo-specificity* | $k_H/k_D$ |
|---|---|---|---|---|---|
| H-abstraction by *t*-BuO· | 1 | 12 | 44 | 0 | 3.7 |
| Singlet ROOC—N insertion | 1 | 10 | 30 | 100 | 1.5 |
| Singlet $CH_2$ insertion | 1 | 1.2 | 1.5 | 100 | 1.3–2 |

* % retention of configuration.

types of bonds, of singlet methylene and that of other reactive intermediates is shown in Table 17.1.

Whereas methylene is extremely reactive, other carbenes show much lower reactivity. Halogens appear to be particularly effective in stabilizing carbenes. As can be seen from data in Table 17.2, the selectivity of insertion into C—H bonds increases with increasing stabilization of the carbene.

**Table 17.2** Selectivity of Carbene Insertion into C—H Bonds

| Carbene | 1° | 2° | 3° |
|---|---|---|---|
| $H_2C$: | 1 | 1.2 | 1.5 |
| $CH_3OOC$—CH: | 1 | 2.3 | 2.9 |
| $(CH_3OOC)_2C$: | 1 | 4 | 13 |
| $C_6H_5$—CH: | 1 | 8.3 | |
| Cl—CH: | 1 | 20 | |
| Br—CH: | 1 | 25 | |

The insertion of carbene into C—H bonds can take place by either a direct, one step mechanism or an abstraction-recombination, two-step process. An experiment to distinguish between these two mechanisms was

$$\rangle C{-}H + CH_2 \rightarrow \left( \rangle C \begin{smallmatrix} H \\ \vdots \\ CH_2 \end{smallmatrix} \right) \rightarrow \rangle C{-}CH_2{-}H$$

$$\rangle C{-}H + CH_2 \rightarrow \rangle C\cdot + \cdot CH_3 \rightarrow \rangle C{-}CH_3$$

designed by Doering.[10] The photolysis of diazomethane in liquid $^{14}C$-labelled isobutene was found to result in only 1% scrambling of label in the product 2-methylbutene-1, from which finding it was concluded that insertion of singlet methylene proceeds via a direct insertion transition state (6) rather than the formation of an intermediate radical pair (7).

$$CH_2N_2 + (CH_3)_2C{=}\overset{*}{C}H_2 \xrightarrow{h\nu} (CH_3)(CH_3{-}CH_2)C{=}\overset{*}{C}H_2\ (99\%) + (CH_3)(CH_2{=})C{-}\overset{*}{C}H_2{-}CH_3\ (1\%)$$

$$H\cdots CH_2{-}C(CH_3){=}\overset{*}{C}H_2 \text{ (bridged by } CH_2) \quad (6) \qquad H_2C{\overset{\cdot}{\frown}}C(CH_3){\frown}CH_2 \quad \cdot CH_3 \quad (7)$$

(C* for $^{14}C$-labelled carbon atom)

The two-step abstraction-recombination mechanism, however, operates in reactions involving triplet carbenes. Triplet diphenylcarbene, for instance, inserts into allylic C—H bonds by this mechanism.

$$Ph_2\,\uparrow\uparrow + (H)(CH_3)C{=}C(CH_3)(H) \rightarrow Ph_2\dot{C}H + (H)(H_2C)\overset{\cdot}{C{\frown}C}(CH_3)(H)$$

$$\rightarrow (H)(Ph_2CH{-}CH_2)C{=}C(CH_3)(H) + (H)(CH_2{=})C{-}CH(CH_3)(CHPh_2)$$

Except for the more reactive carbenes the insertion into unactivated C—H bonds is a relatively rare occurrence. Addition to unsaturated bonds, on the other hand, represents a lower energy pathway. The addition of singlet carbene to olefins involves the attack of the vacant *p* orbital of the carbene on the olefinic electrons. Evidence available at present indicates that the mechanism is a one-step or direct addition. A completely symmetrical transition state in the addition of carbene to olefin is forbidden on orbital symmetry grounds but the unsymmetrical addition involving orbital rehybridization in the transition state is allowed.[11] This one-step or direct addition mechanism is in agreement with the observation that additions of

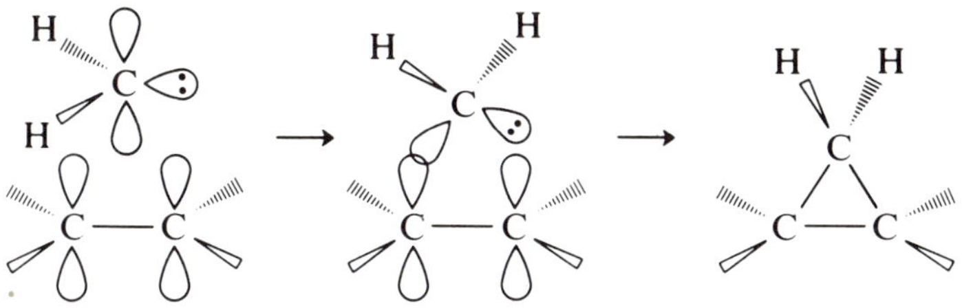

singlet carbene always proceed with complete stereospecificity, i.e., the original configuration of the substituents in the olefin is preserved in the cyclopropane produced. The stereospecific addition of dihalocarbene to the isomers of butene-2 is shown below.

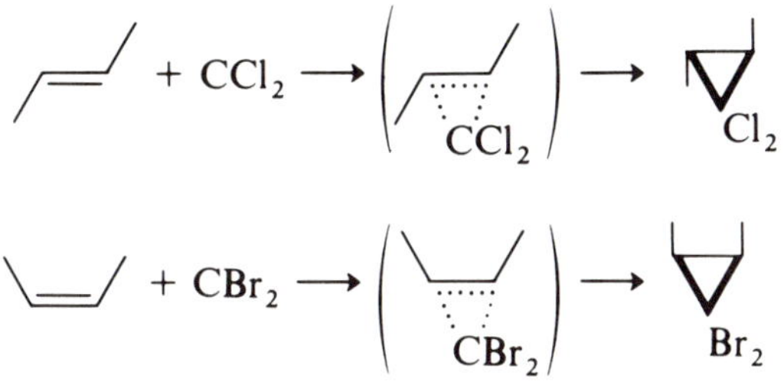

$$+ CCl_2 \longrightarrow \left( CCl_2 \right) \longrightarrow Cl_2$$

$$+ CBr_2 \longrightarrow \left( CBr_2 \right) \longrightarrow Br_2$$

Singlet carbene is essentially an electrophilic species. The classical generation of dichlorocarbene and its reaction with the phenoxide ion in the

$$Cl_3CH + OH^- \rightleftarrows Cl_3C^- \qquad \text{fast}$$

$$Cl_3C^- \rightarrow Cl_2C{:} + Cl^- \qquad \text{slow}$$

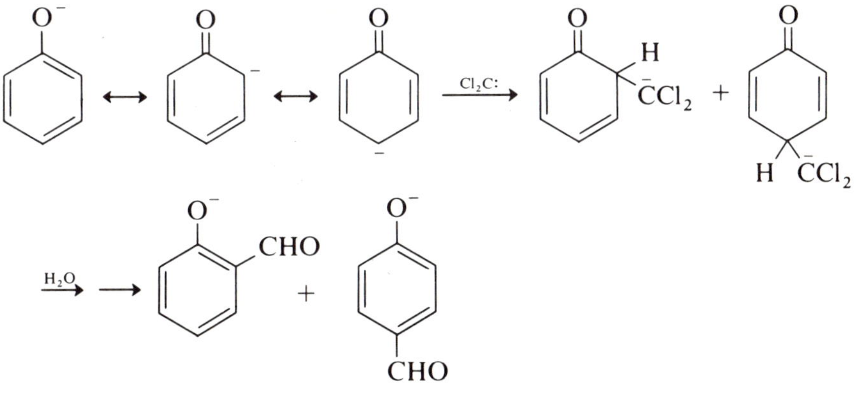

$$\xrightarrow{Cl_2C:} \quad CCl_2 \quad + \quad H \; \bar{C}Cl_2$$

$$\xrightarrow{H_2O} \longrightarrow \quad CHO \quad + \quad CHO$$

**Table 17.3** Relative Rates of Addition of Various Carbenes to Olefins

| Olefin | Carbene | | | | |
|---|---|---|---|---|---|
| | $H_2C:$ | $C_6H_5CH:$ | $ClCH:$ | $Cl_2C:$ | $Br_2C:$ |
| 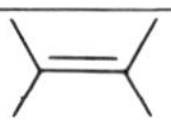 | 1.55 | | 1.10 | 43 | 3.5 |
| 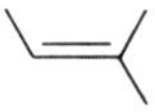 | 1.52 | 3.3 | 1.08 | 19 | 3.2 |
| 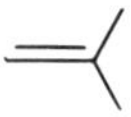 | 1.43 | 0.91 | 0.92 | 6.7 | 1.00 |
|  | 0.99 | 1.84 | 0.91 | 1.5 | |
| 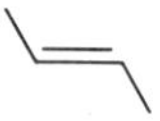 | 1.00 | 1.00 | 1.00 | 1.00 | |
|  | 1.17 | 1.0 | 0.71 | 0.073 | |

Riemer-Teiman reaction demonstrates this point. The relative rates of addition to olefins depend also on the nucleophilicity of the olefins. Table 17.3 shows the reactivity of various olefins towards a representative group of carbenes. Among the olefins the more alkylated ones, being relatively electron-rich, react faster except where a large unfavourable steric effect is present. Among the carbenes, it is noteworthy that, while singlet methylene adds to olefins rather indiscriminately, carbenes carrying substituents capable of conferring stability through resonance exhibit markedly higher selectivity. Thus the relatively high selectivity of dichlorocarbene indicates that halogens can stabilize carbenes by resonance, probably as follows:

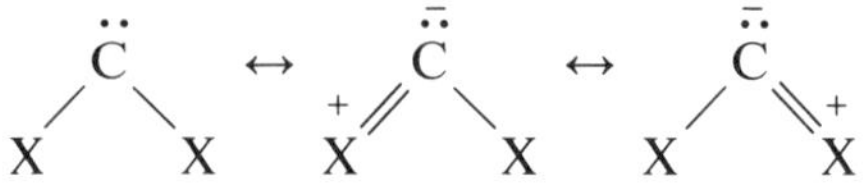

$$X{-}\ddot{C}{-}X \leftrightarrow \overset{+}{X}{=}\bar{\ddot{C}}{-}X \leftrightarrow X{-}\bar{\ddot{C}}{=}\overset{+}{X}$$

Such stabilizing capacities of substituent groups appear to be approximately of the order F > Cl > Br > $C_6H_5$ > COOR.

The importance of polar effects in carbene additions may be studied using Hammett type linear free energy relationships. The reaction of dichlorocarbene with substituted α-methylstyrenes gave a correlation with $\sigma^+$, $\rho$ being $-0.378$ (Sadler).[12] This is in agreement with the electrophilic character of the reacting species. However, except for reactions in which the more strongly electrophilic carbenes, such as dichloromethylene, take part, one may not expect strongly polar transition states.

It is sometimes difficult to determine whether a free carbene species is a singlet or a triplet. Towards a solution of this problem, Skell[13] put forward the hypothesis that triplet carbene can add non-stereospecifically to olefins whereas singlet carbene adds stereospecifically. Non-stereospecificity of addition is due to the fact that a triplet intermediate (8) has to undergo spin inversion to get to a ground state product but the rate of spin inversion ($k_i$) may be comparable to the rate of bond rotation ($k_r$) of the intermediate. It is noted that Skell's hypothesis only states that a non-stereospecific product is due to a triplet carbene but does not mean that a stereospecific product cannot arise from such a species. Further, when a mixture of stereospecific and

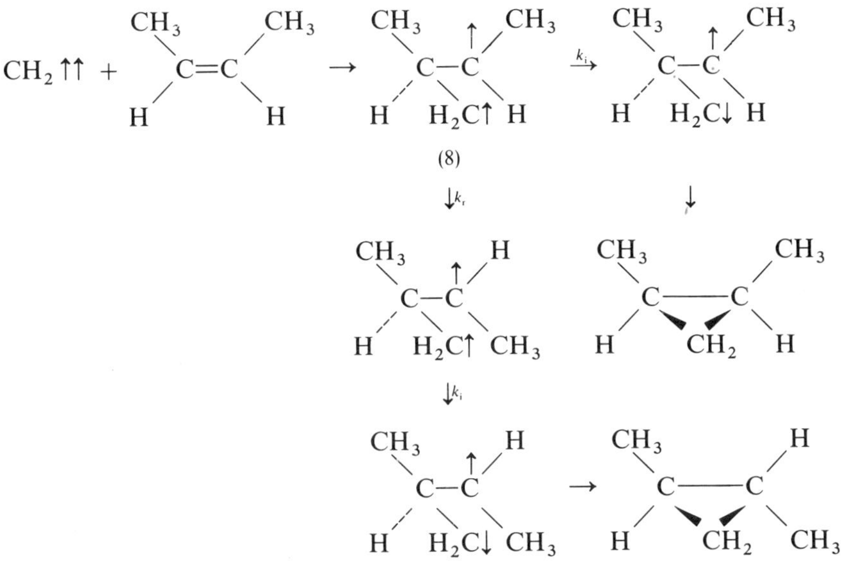

(8)

non-stereospecific addition products appear, a mixture of singlet and triplet carbenes may be involved, not just the triplet. The photolysis of 9-diazofluorene provides an example in which both triplet and singlet carbenes are formed; addition of the carbene to *cis* butene gives both *cis* and *trans* cyclopropanes. The yield of the latter may be decreased by the presence of a triplet scavenger (butadiene or oxygen), or it may be increased by a suitable

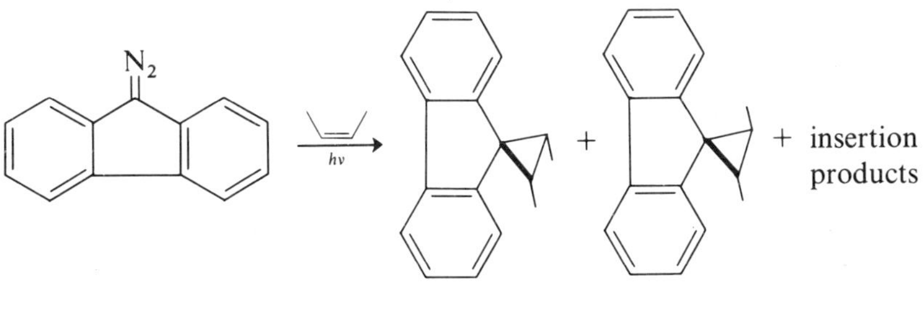

+ insertion products

inert solvent (hexafluorobenzene) which allows for more collisional conversion of the initially formed singlet species to the triplet.

In the gas phase the chemistry of methylene ($CH_2$:) is more complicated than in the liquid phase. This is due to the fact that the singlet methylene initially formed can undergo spin relaxation to the lower energy triplet before reaction, as a result of the much lower frequency of collisions in the gas phase than in the liquid phase. Furthermore, the addition of methylene to olefin is highly exothermic ($\Delta H = -86$ kcal mole$^{-1}$) and invariably leads to a vibrationally excited or 'hot' molecule which can isomerize readily before collisional deactivation occurs. The photolysis of diazomethane in *cis* butene gas leads to a number of products:

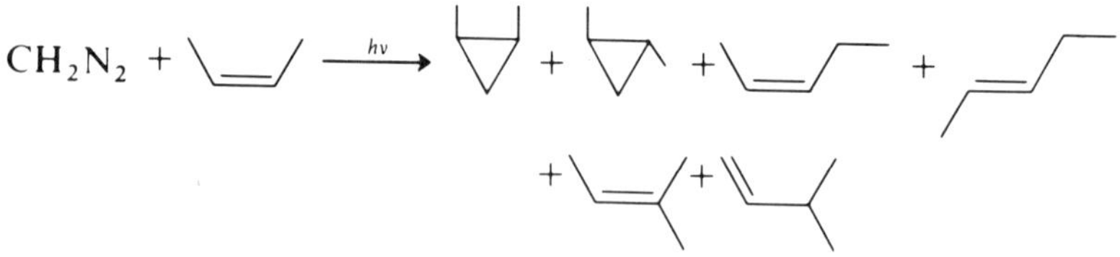


The formation of these products has been explained as due to both the reactions of a mixture of singlet and triplet methylenes as well as the isomerization of 'hot' cyclopropane molecules initially formed from singlet carbene addition. The formation of the non-stereospecific product, *trans* dimethylcyclopropane, may be reduced slightly by increasing the pressure of *cis* butene. This indicates that the 'hot' *cis* dimethylcyclopropane formed from the addition of singlet $CH_2$ to *cis* butene may be collisionally deactivated before isomerization occurs. On the other hand, the continued addition of inert gases, such as argon or nitrogen, to the system results in the non-stereospecific addition being increased. This is interpreted as due to the increased formation of triplet $CH_2$ as the presence of the inert gas leads to a larger number of fruitless collisions of singlet $CH_2$ with surrounding molecules with consequent rise in decay to the triplet state. On the other hand oxygen, a ground state triplet which effectively scavenges triplet $CH_2$, reduces non-stereospecific addition.

Carbenoid species (5), in general, are less reactive than the corresponding free carbenes. Being less reactive they show more selectivity in their additions to unsaturated bonds. Insertion reactions of carbenoids are rare and when they occur are usually intramolecular in nature. The better studied carbenoids are the α-halo-organolithiums and α-halo-organozinc halides. Although these compounds are usually formulated simply as (5) they are probably dimers, tetramers or ion clusters, rather than monomers. Their addition to olefins is completely stereospecific and generally much more selective than the free carbenes (see Table 17.4). The relative reactivities

**Table 17.4** Relative Rates of Carbenoid Olefin Additions

| Olefin | $ICH_2ZnI$ | $ArCHBrLi$[a] | $ArCHBrZnBr$[b] | $Cl_2CHLi$[c] |
|---|---|---|---|---|
| | 3.1 | | | 6.2 |
| | 5.2 | 1.50 | 2.70 | 4.0 |
| | 6.0[d] | 1.35 | 2.54 | 2.2 |
| | 2.0[e] | 1.35 | 2.11 | 2.0 |
| | 1.0[f] | 1.0 | 1.0 | 1.0 |
| | 0.862[g] | 0.72 | 0.44 | 0.51 |
| | 2.4 | | | 1.3 |

a, $p$-$CH_3$—$C_6H_4$—$CHBr_2$ + $CH_3Li$; b, $p$-$CH_3$—$C_6H_4$—$CHN_2$ + $ZnBr_2$
c, $CH_2Cl_2$ + RLi; d, 2-Methyl-1-butene; e, *cis*-3-hexene; f, *trans*-3-hexene;
g, 1-hexene.

appear to depend on the nucleophilicity of the olefins as well as on steric factors.

The mechanism of a carbenoid addition may be written as a direct, one-step addition with the olefin acting as a nucleophile displacing MX:

$$>C=C< \; + \; \underset{R'}{\overset{R}{>}}C\underset{X}{\overset{M}{<}} \;\searrow\; \left( \text{C}\cdots\text{C}(R)(R')(M)\cdots X \right) \longrightarrow \text{cyclopropane}(C, C, C(R)(R')) + MX$$

This mechanism is in harmony with the stereospecificity of the reaction. Steric effects can also influence the course of carbenoid additions. Remarkable stereoselectivity has been observed for several carbenoid additions (Closs, *et al.*).[5a,14] Thus arylcarbenoids and chlorocarbenoids add to *cis* butene to give predominantly the thermodynamically less stable *syn* cyclopropane isomer. Still greater stereoselectivity has been observed in the reactions of zinc carbenoids. These reactions are depicted as follows:

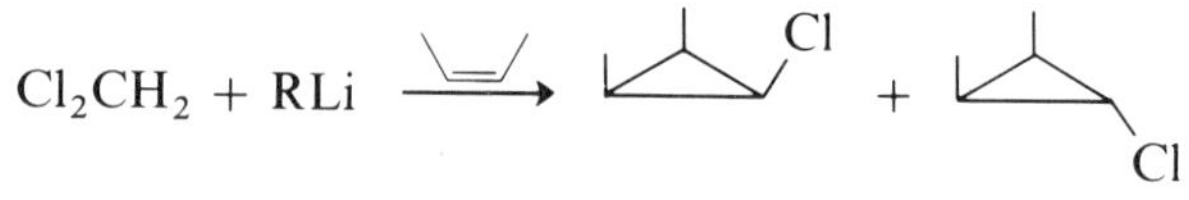

$$Cl_2CH_2 + RLi \xrightarrow{\text{cis-2-butene}} \text{syn-Cl cyclopropane} + \text{anti-Cl cyclopropane}$$

*syn* *anti*

*syn/anti* isomer ratio = 5.5

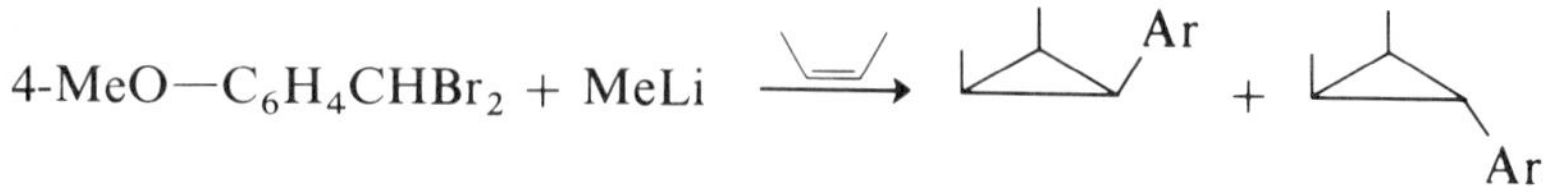

$$4\text{-MeO}{-}C_6H_4CHBr_2 + MeLi \xrightarrow{\text{cis-2-butene}} \text{syn-Ar} + \text{anti-Ar}$$

*syn/anti* isomer ratio = 8.1

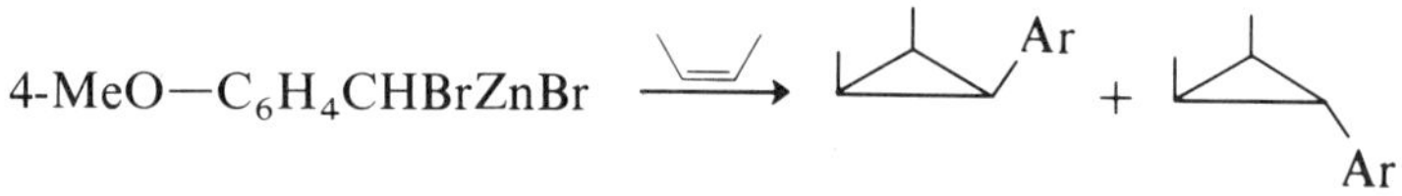

$$4\text{-MeO}{-}C_6H_4CHBrZnBr \xrightarrow{\text{cis-2-butene}} \text{syn-Ar} + \text{anti-Ar}$$

*syn/anti* isomer ratio = 53

The observed *syn* stereoselectivity of carbenoid addition has been explained by a transition state in which the aryl $\pi$-electrons interact by Van der Waals and London dispersion forces with the *cis* olefinic alkyl substituents (9).

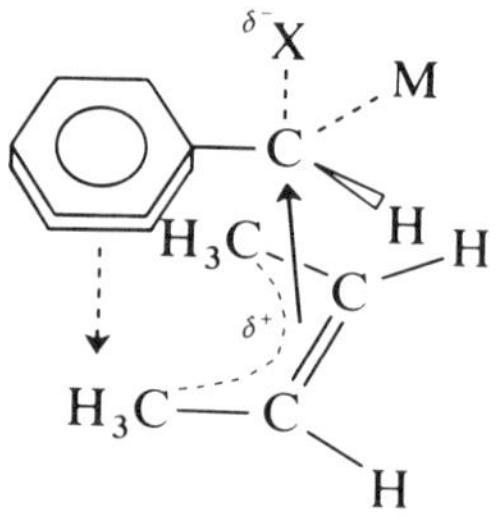

(9)

## 17.4 Nitrenes and other reactive intermediates

As carbenes are divalent carbon species, nitrenes may be referred to as monovalent nitrogen intermediates. Like carbenes both singlet and triplet states are possible for nitrenes. Nitrenes may be generated by photolysis or thermolysis of azides or by $\alpha$-elimination.[15] Nitrenes generated by these methods are initially in the singlet state but partially decay to the triplet

$$EtO_2C{-}N{=}\overset{+}{N}{=}\overset{-}{N} \xrightarrow{h\nu} EtO_2C{-}N + N_2$$

$$C_6H_5{-}N{=}\overset{+}{N}{=}\overset{-}{N} \xrightarrow{\Delta H} C_6H_5{-}N + N_2$$

$$EtO_2C{-}NHOSO_2Ar + Et_3N \rightleftarrows EtO_2C{-}\overset{-}{N}{-}OSO_2Ar \rightarrow EtO_2{-}N + ArSO_3^-$$

state before reaction. The addition of ethoxycarbonylnitrene to *cis* methylisopropylethylene is non-stereospecific and can be rationalized using Skell's hypothesis. The percentage of non-stereospecific addition product increases

$$EtO_2C{-}N_3 \rightarrow EtO_2C{-}N\,\uparrow\downarrow \xrightarrow{\text{Me–CH=CH–}i\text{-Pr}} \text{(Me)(H)C–C(H)(}i\text{-Pr) aziridine, N–CO}_2\text{Et} \quad (86\%)$$

$$EtO_2C{-}N_3 \rightarrow EtO_2C{-}N\,\uparrow\uparrow \xrightarrow{\text{Me–CH=CH–}i\text{-Pr}} \text{(Me)(H)}\dot{C}\text{–C(H)(}i\text{-Pr)–}\dot{N}\text{CO}_2\text{Et} \rightarrow (86\%\ \text{product})$$

$$\text{(Me)(H)}\dot{C}\text{–C(H)(}i\text{-Pr)–}\dot{N}\text{CO}_2\text{Et} \;\uparrow\downarrow\; \text{(H)(Me)}\dot{C}\text{–C(H)(}i\text{-Pr)–}\dot{N}\text{CO}_2\text{Et} \longrightarrow \text{(H)(Me)C–C(}i\text{-Pr)(H) aziridine, N–CO}_2\text{Et} \quad (14\%)$$

with the addition of methylene chloride as solvent, which is in agreement with the view that the initially formed singlet nitrene is collisionally deactivated to the triplet.

The insertion of singlet ethoxycarbonylnitrene into C—H bonds proceeds with retention of configuration.[16] This insertion may be viewed as a direct

$$(EtCH_2)(Et)(Me)\overset{*}{C}{-}H \xrightarrow{EtO_2C{-}N\,\uparrow\downarrow} \left( (EtCH_2)(Et)(Me)C \cdots H \cdots NCO_2Et \right) \longrightarrow (EtCH_2)(Et)(Me)\overset{*}{C}{-}NHCO_2Et$$

$$(Ph)(Et)(Me)\overset{*}{C}{-}H \xrightarrow{Ph{-}N\,\uparrow\uparrow} (Ph)(Et)(Me)C\cdot + Ph\dot{N}H \longrightarrow (Ph)(Et)(Me)C{-}NHPh$$

40% retention

insertion proceeding through a three-centre transition state. The insertion by a triplet nitrene (e.g., phenylnitrene) however, leads to partial racemization,[17] which indicates that an abstraction-recombination process has taken place.

In the study of carbenes atomic carbon is also of interest. Although the electronic states of atomic carbon are well understood the generation of atomic carbon and a study of its reactions have presented a number of difficulties. Skell, *et al.*,[18] have recently succeeded in producing atomic carbon atoms from a low intensity carbon arc under high vacuum. By this method the following three reactive species have been investigated: triatomic carbon (10), singlet atomic carbon (11), and triplet atomic carbon (12). Triatomic carbon, a dicarbene, forms a major component of carbon vapour.

:C=C=C: (10) :C ⇅ (11) :Ċ↑ (12)

Its reaction with olefins deposited as a thin film at −196 °C leads to cyclopropane adducts. In the same way singlet atomic carbon was shown to react with olefins and hydrocarbons by addition and insertion. If the

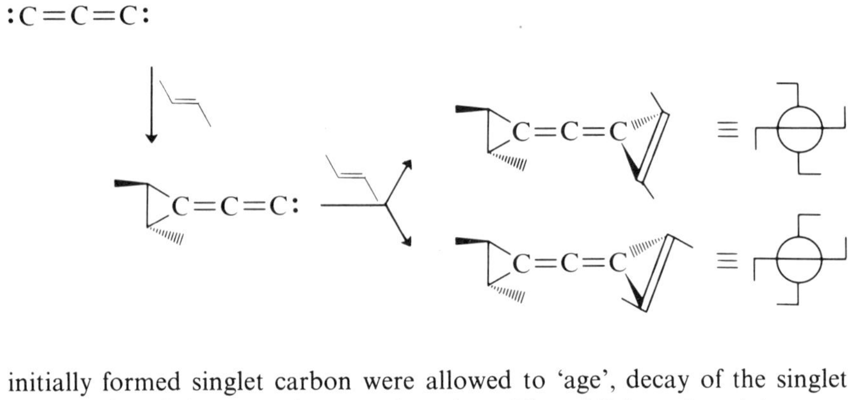

:C=C=C:

initially formed singlet carbon were allowed to 'age', decay of the singlet state to the triplet ground state takes place. The addition of a triplet state carbon atom to olefin molecules gives spiropentanes.

The chemistry of triatomic carbon and carbon atoms shows a close resemblance to divalent carbon intermediates. Very little is known about the reactions of monovalent carbon intermediates or carbynes. Recently the photolysis of diethyl mercurybisdiazoacetate, $Hg(N_2CCO_2Et)_2$, was reported to give ethoxycarbonylcarbyne (:Ċ—$CO_2$Et), which reacts with cyclohexene to give a number of products.[19] The first step of the reaction

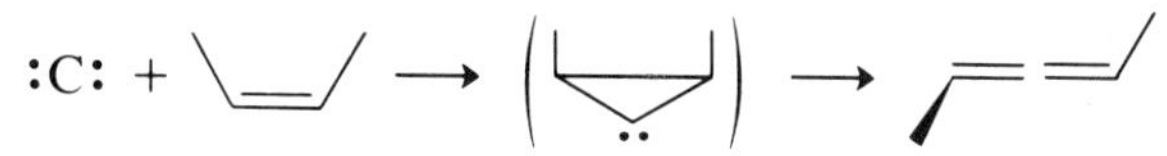

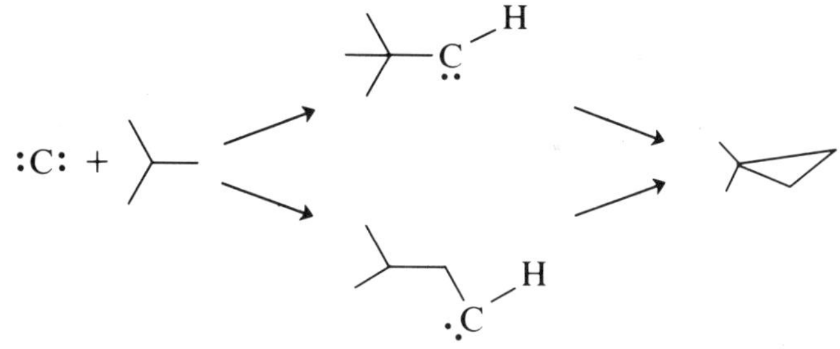

:C: +

:C: +

involves an insertion or addition reaction, giving radical intermediates which abstract allylic protons from cyclohexene to give the products.

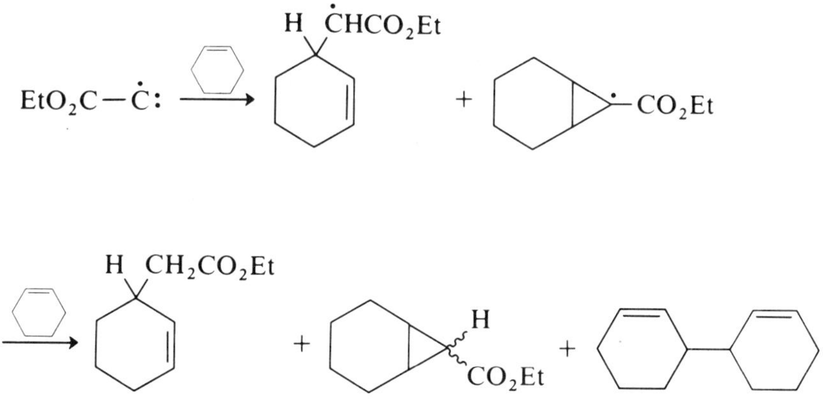

## References


1*a* G. L. Closs, Structures of Carbenes and the Stereochemistry of Carbene Additions to Olefins, in *Topics in Stereochemistry*, vol. 3, Interscience, New York, N.Y., 1968.

*b* R. A. Moss, Carbene Chemistry, *Chem. Engng. News*, 1969, June 30, p. 50.

*c* D. Bethell, Structure and Mechanism in Carbene Chemistry, *Adv. Phys. Org. Chem.*, 1969, **7**, 153–209.

*d* W. Kirmse, *Carbene Chemistry*, Academic Press, New York, N.Y., 2nd Edition, 1971.

*e* W. E. Parham and E. E. Schweizer, Halocyclopropanes from Haloforms, in *Organic Reactions*, Vol. 13, John Wiley & Sons, Inc., New York, N.Y., 1963.

2 G. Herzberg, *Proc. R. Soc. (London)*, 1961, **A262**, 291.

3*a* R. W. Brandon, G. L. Closs, and C. A. Hutchinson, *J. chem. Phys.*, 1962, **37**, 1878.

*b* A. M. Trozzolo, R. W. Murray, and E. Wasserman, *J. Am. chem. Soc.*, 1962, **84**, 4990.

4 P. S. Skell and M. S. Cholod, *J. Am. chem. Soc.*, 1969, **91**, 6035.

5*a* G. L. CLOSS and R. A. MOSS, *J. Am. chem. Soc.*, 1964, **86**, 4042.
*b* G. KOBRICH, *Angew. Chem., Int. Ed.*, 1967, **6**, 41.
6 J. HINE, *Divalent Carbon*, Ronald Press Co., New York, N.Y., 1964.
7*a* H. E. SIMMONS and R. D. SMITH, *J. Am. chem. Soc.*, 1958, **80**, 5323.
*b* G. WITTIG and K. SCHWARZENBACH, *Angew. Chem.*, 1959, **71**, 652.
*c* E. P. BLANCHARD and H. E. SIMMONS, *J. Am. chem. Soc.*, 1964, **86**, 1337.
8 D. SEYFERTH, J. M. BURLITCH, R. J. MINASZ, J. Y. P. MUI, H. D. SIMMONS, JR., A. J. H. TREIBER, and S. R. DOWD, *J. Am. chem. Soc.*, 1965, **87**, 4259.
9 W. v. E. DOERING, R. G. BUTTERY, R. G. LAUGHLIN, and N. CHAUDHURI, *J. Am. chem. Soc.*, 1956, **78**, 3224.
10 W. v. E. DOERING and H. PRINZBACH, *Tetrahedron*, 1959, **6**, 24.
11 R. HOFFMANN, *J. Am. chem. Soc.*, 1968, **90**, 1475.
12 I. H. SADLER, *J. chem. Soc. (B)*, 1969, 1024.
13*a* P. S. SKELL and R. C. WOODWORTH, *J. Am. chem. Soc.*, 1956, **78**, 4496.
*b* S. H. GOH, *Chem. Commun.*, 1972, 512.
14*a* S. H. GOH, L. E. CLOSS, and G. L. CLOSS, *J. org. Chem.*, 1969, **34**, 25.
*b* C. D. POULTER, E. C. FRIEDRICH, and S. WINSTEIN, *J. Am. chem. Soc.*, 1969, **91**, 6892.
*c* L. Y. GOH and S. H. GOH, *J. Organometal. Chem.*, 1970, **23**, 5.
15 J. S. MCCONAGHY, JR. and W. LWOWSKI, *J. Am. chem. Soc.*, 1967, **89**, 4450.
16 G. SMOLINSKY and B. I. FEUER, *J. Am. chem. Soc.*, 1964, **86**, 3085.
17 J. H. HALL, J. W. HILL, and J. M. FARGHER, *J. Am. chem. Soc.*, 1968, **90**, 5313.
18*a* P. S. SKELL, L. D. WESCOTT, JR., J. P. GOLSTEIN, and R. R. ENGEL, *J. Am. chem. Soc.*, 1965, **87**, 2829.
*b* P. S. SKELL and R. R. ENGEL, *J. Am. chem. Soc.*, 1967, **89**, 2912.
19 THAP DOMINH, H. E. GUNNING, and O. P. STRAUSZ, *J. Am. chem. Soc.*, 1967, **89**, 6785.

# Author Index

Greene, F. D., 48, 49, 51, 189, 192, 199, 203

# Subject Index